Blockchain Technology in Internet of Things

Liehuang Zhu • Keke Gai • Meng Li

Blockchain Technology in Internet of Things

 Springer

Liehuang Zhu
School of Computer Science and
Technology
Beijing Institute of Technology
Beijing, China

Keke Gai
School of Computer Science and
Technology
Beijing Institute of Technology
Beijing, China

Meng Li
College of Computer Science and
Information Engineering
Hefei University of Technology
Hefei, China

ISBN 978-3-030-21768-6   ISBN 978-3-030-21766-2   (eBook)
https://doi.org/10.1007/978-3-030-21766-2

This Springer imprint is published by the registered company Springer Nature Switzerland AG.
The registered company address is: Gewerbestrasse 11, 6330 Cham, Switzerland

Dedication

We pay our highest respect and gratitude to the editors for their support in publishing this book and thank all reviewers and peers for their precious advice in bettering this book. We also thank the School of Computer Science and Technology in Beijing Institute of Technology and all our colleagues in the lab for their support and help in completing this book. We further appreciate our families for their unconditional love. We still remember the holidays and vacations when we were finishing this book while not keeping our families company.

Dr. Zhu would like to thank his wife, Fang Guan; daughter, Jie Zhu; mother, Lingying Zeng; brothers, Liehui Zhu; and many other relatives for their continuous love, support, trust, and encouragement throughout his life. Without them, none of this would have happened.

Dr. Gai dedicates this work to his parents, father Jinchun Gai and mother Tianmei Li, who have brought him up and sacrificed so much, as well as his wife, Xiaotong Sun. He could never have done this without his parents' and wife's love, support, and constant encouragement. A sincere appreciation to all Keke's family members for their continuous love.

Meng Li would like to thank his father, Ruxue Li, and his mother, Hui Wang, for their endless love and care. He gives his gratitude to his doctoral supervisor and co-supervisors, Liehuang Zhu, Zijian Zhang, and Xiaodong Lin. Finally, he is grateful for everything he has experienced, acquired, and expected in the Beijing Institute of Technology.

Foreword

This book focuses on picturing B-IoT techniques from a few perspectives, which are architecture, key technologies, security and privacy, service models and framework, practical use cases, etc. The main contents of this book are derived from the most updated technical achievements or breakthroughs in the field. A number of representative IoT service offerings will be covered by this book, such as vehicular networks, document sharing system, and telehealth. Both theoretical and practical contents will be involved in this book in order to assist the readers to have a comprehensive and deep understanding of the mechanism of using blockchain for powering up IoT systems.

This book facilitates students and learners to acquire some basic knowledge of IoT and blockchain and supports both practitioners and scholars to find some new insights in B-IoT systems.

Preface

The blockchain-enabled Internet of Things (B-IoT) is deemed a novel technical alternative that provides network-based services with additional functionalities, benefits, and implementations in terms of decentralization, immutability, and auditability. Towards the enhanced secure and privacy-preserving Internet of Things (IoT), this book introduces few significant aspects of B-IoT, which includes the fundamental knowledge of both blockchain and IoT, state-of-the-art reviews of B-IoT applications, crucial components in the B-IoT system and the model design, and future development potentials and trends.

IoT technologies and services, e.g., cloud data storage technologies and vehicular services, play important roles in wireless technology developments. On the other side, blockchain technologies are being adopted in a variety of academic societies and professional realms due to its promising characteristics. It is observable that the research and development on integrating these two technologies will provide critical thinking and solid references for contemporary and future network-relevant solutions.

This book focuses on picturing B-IoT techniques from a few perspectives, which are architecture, key technologies, security and privacy, service models and framework, practical use cases, etc. The main contents of this book derive from most updated technical achievements or breakthroughs in the field. A number of representative IoT service offerings will be covered by this book, such as vehicular networks, document sharing system, and telehealth. Both theoretical and practical contents will be involved in this book in order to assist the readers to have a comprehensive and deep understanding of the mechanism of using blockchain for powering up IoT systems.

This book facilitates students and learners to acquire some basic knowledge of IoT and blockchain and supports both practitioners and scholars to find some new insights in B-IoT systems.

Beijing, China   Liehuang Zhu
Beijing, China   Keke Gai
Hefei, China   Meng Li

Acknowledgments

The authors in this book intend to express their highest and truest gratitude to all the participants who contribute and support this work. All the supports, advice, and encouragement made by peer experts, scholars, students, and the lab in BIT are remarkably meaningful and invaluable for the accomplishment of this book. The authors sincerely appreciate all the individuals and organizations who reinforce and raise the quality of this book.

Beijing, China   Liehuang Zhu
Beijing, China   Keke Gai
Hefei, China   Meng Li
February 2019

Contents

About the Authors

Liehuang Zhu is a Professor and the Vice Dean in the School of Computer Science and Technology at Beijing Institute of Technology. He is selected into the Program for New Century Excellent Talents in the university from the Ministry of Education, P. R. China. He has published 50+ journal papers and 40+ conference papers in recent years, including IEEE TDSC, *IEEE TIFS, IEEE Communications Magazine, IEEE Wireless Communications, IEEE IoT Journal, IEEE Network, IEEE TSG, IEEE TVT, IEEE Access, Information Sciences*, IEEE/ACM IWQoS, IEEE IPCCC, and IEEE GLOBECOM. He has served as the Chair in SmartBlock 2018, and the Program Committee Chair in BcADS 2019, MSN 2017, InTrust 2014, and InTrust 2011. He is an Associate Editor for the IEEE Transactions on Vehicular Technology. He was a Guest Editor for the *IEEE Wireless Communications Magazine* and IEEE Transactions on Industrial Informatics. He has been granted three best paper awards in IEEE/ACM conferences, including IEEE TrustCom 2018, IEEE/ACM IWQoS 2017, and IEEE IPCCC 2014. He has been awarded as an Excellent Advisor in China Institute of Communications Excellent Doctoral Dissertation and China National College Student Information Security Contest. His research interests include cryptographic algorithms and secure protocols, Internet of Things security, cloud computing security, big data privacy, mobile and Internet security, and trusted computing.

Keke Gai received a Ph.D. degree in Computer Science from the Department of Computer Science at Pace University, New York, USA. He also holds degrees from Nanjing University of Science and Technology (BEng), The University of British Columbia (MET), and Lawrence Technological University (MBA and MS). He is currently an Associate Professor in the School of Computer Science and Technology at Beijing Institute of Technology, Beijing, China. He has published more than 100 peer-reviewed journals or conference papers in recent years, including 40+ journal papers (including ACM/IEEE Transactions) and 50+ conference papers. He has been granted five IEEE Best Paper Awards (IEEE TrustCom 2018, IEEE HPCC 2018, IEEE SSC 2016, IEEE CSCloud 2015, IEEE BigDataSecurity 2015) and two IEEE Best Student Paper Awards (IEEE HPCC 2016, IEEE SmartCloud 2016) by IEEE conferences in recent years. His paper about edge computing has been granted the Best Paper of the Year by *Journal of Network and Computer Applications* (JNCA) in 2018. He is involved in a number of professional/academic associations, including ACM and IEEE. Currently, he is serving as the Secretary/Treasurer of the IEEE STC (Special Technical Community) in Smart Computing at the IEEE Computer Society. He has worked for a few Fortune 500 enterprises, including SINOPEC and GE Capital. His research interests include blockchain, cyber security, cloud computing, edge computing, and combinatorial optimization. He also served as Program Chairs in several international conferences, such as EUC 2019, BSCI 2019, SmartCom 2019, SmartBlock 2018, etc.

Meng Li is now an Associate Researcher in the the College of Computer Science and Information Engineering at the Hefei University of Technology. He received his Ph.D. degree from the School of Computer Science and Technology at Beijing Institute of Technology. He was a visiting Ph.D. Student in Wilfrid Laurier University. He has published over 20 journal and conference papers including IEEE TDSC, *IEEE IoT Journal,* IEEE TSC, *IEEE Wireless Communications, IEEE Communications Magazine,* and IEEE IPCCC. He has received the National Graduate Student Scholarship in 2011 and has been sponsored by China Scholarship Council High-Level University Graduate Student Government-Sponsored Program and Beijing Institute of Technology High-Level Ph.D. Dissertation Nursery Funding in 2017. He received the National Ph.D. Graduate Scholarship in 2018. His research interests include applied cryptography, security and privacy, vehicular networks, fog computing, and blockchain.

Acronyms

This part provides audiences with a list of acronyms. The abbreviations in the list are unified in this book. Most selected abbreviations are broadly accepted concepts in computer science domain.

BCDM   Blockchain-Based controllable data management
B-IoT   Blockchain-Enabled Internet of Things
BLS   Bone-Lynn-Shacham
BTC   Bitcoin
CB   Consortium blockchain
CC   Cloud computing
CDH   Computational Diffie-Hellman
CDP   Cloud data preservation
CMA   Chosen-Message Attack
CP   Coin pool
CPA   Chosen-Plaintext Attack
CS   Cloud server
DMS   Data management system
DPOS   Delegated Proof of Stake
ECDSA   Elliptic Curve Digital Signature Algorithm
ETH   Ether
EV   Electric vehicle
EVM   Ethereum Virtual Machine
FC   Fog computing
IoT   Internet of Things
KMA   Known-Message Attack
KPA   Known-Plaintext Attack
LA   Local aggregator
LBS   Location-based services
LPPM   Location privacy protection mechanism
NFC   Near Field Communication
OBU   On-board unit

PCA	Pseudonym certification authority
PCS	Privacy-Preserving carpooling services
POI	Points-of-Interest
PoS	Proof-of-Stake
PoW	Proof-of-Work
PPT	Probabilistic polynomial-time
PRHS	Privacy-Preserving ride-hailing services
PS	Preservation Submission
PV	Primitiveness Verification
P2P	Peer-to-peer
RA	Registration authority
RFID	Radio Frequency Identification
RHS	Ride-hailing service
RMA	Random-Message Attack
RSU	Road-side unit
SA	System administrator
SCTS	Supply chain traceability system
SG	Smart grid
SHN	Smart home network
SM	Smart meter
SP	Service provider
StaaS	Storage-as-a-Service
TA	Trusted authority
TD	Truth discovery
TP	Transaction pool
Tx	Transaction
UTXO	Unspent Transaction Output
VET	Vehicle electricity transaction
zk-SNARK	Succinct non-interactive zero-knowledge proof

Part I
Basic Concepts and Mechanisms of Blockchain in Internet of Things

Chapter 1
Introduction

1.1 Overview

Blockchain has been firstly introduced to public along with the emergence of Bitcoin in 2008 [1]. It is widely known as a decentralized and tamper-resistant ledger technique that is maintained by a group of users with specific purpose(s). Considering the scenario of Bitcoin, such a ledger keeps records of all transactions, e.g., financial deals [2–4], supply chain information, and copyright ownership. After the smart contract was introduced by Ethereum, blockchain was considered a breakthrough technical concept that fitted in a broad scope of implications and applications [5]. We consider blockchain to be a key term throughout this book.

Meanwhile, in the era of the Internet, Internet of Things (IoT) [6] is also enabled by a group of rapid advancements, such as mobile sensors, high-capacity computation, and communication protocols [7–9]. The collaboration of devices are deemed the "things" by which environmental data are sensed and gathered for the purpose of controlling data center or service offerings. Some of the IoT applications are represented by wireless sensor networks, vehicular networks, e-healthcare, and cloud storage services. As an emerging technical subject, IoT is bridging up two worlds, the physical and information worlds, by offering a great range of web-based services [10, 11]. In this book, IoT is a main platform in which blockchain techniques are applied.

Literally speaking, many attempts from both academia and industry have shown a great effort in integrating blockchain techniques with mature IoT services in order to facilitate blockchain-enabled IoT (B-IoT) systems. An example is offering blockchain-enabled cloud storage services [12, 13] where users' digital files are preserved in a cloud server [14] and the hash values of files are recorded on the blockchain. Another example is the blockchain-enabled carpooling service, in which passengers and drivers are matched according to their requests and their carpooling records are stored on blockchain.

Besides functionality brought by integrating blockchain with IoT, security and privacy concerns also are raised in IoT systems for the fact that data-in-transmit in IoT systems contains confidential/sensitive information, e.g., driving destinations, and treatment records [15, 16]. In general, on the one hand, security mainly covers a few aspects, including confidentiality, integrity, authentication, etc.; on the other hand, privacy concerns contain identity, location, trajectory, report, query, etc. It is observable that an adoptable IoT system shall pay a great attention to security and privacy issues.

The concept of the B-IoT system is a novel technical alternative, which provides IoT services with decentralization, immutability, and auditability. This book concentrates on the enhanced B-IoT system in security and privacy-preserving as well as presents a few latest viewpoints/approaches of B-IoT applications. The fundamental knowledge of blockchain and IoT also is covered in this book, too.

1.2 Blockchain

The concept of blockchain often comes with Bitcoin so that this book starts introducing the concept from the scenario of Bitcoin. In the scenario, blockchain is considered a distributed ledger, i.e., a chain of blocks, which stores transactions by packing them into blocks in a chain. The chain links blocks with a cryptographic hash chain, which keeps growing as long as new blocks are created and maintained. It is a technique to enable a type of modern financial systems that tracks transactions/assets without needs of the centralized party.

One of the representative characteristics of blockchain, normally speaking, is decentralized computing with a tamper-resistant feature. The reason of having this characteristic is that no central authority is configured in a blockchain system. A blockchain system is maintained by communications between miners in a distributed network. Data stored in blockchain cannot be falsified once the data are written and stored in a block. In addition, blockchain somehow provides anonymity in a sense where miners can choose multiple public keys to generate identities and participate in blockchain mining. In Bitcoin, mining, e.g., proof-of-work (PoW) and proof-of-stake (PoS), is a way of forming consensus even though an untrustworthy network maybe adopted. A winning miner is periodically selected to create the succeeding block. In order to encourage miners to keep mining, a few incentive mechanisms are designed, such as Bitcoin and Ether.

Basically, blockchain systems can be categorized into three fundamental types, involving public, consortium, and private blockchain systems. These three types of blockchain systems are used in various implementation scenarios, depending on network scale, miner invitation rule, and security level. In general, public blockchain offers a pure decentralized computing environment; thus, public blockchain is extensively implemented in general-purpose financial systems, e.g., Bitcoin blockchain. It has a low requirement towards miner. Differently, private blockchain resides within a single institution, even though it abandons a full decentralized computing setting. It has been widely believed that private blockchain is a type of centralized computing approach. Finally, consortium blockchain fits in a compact cooperation between enterprises/organizations [17], as it positions the mechanism between public and private blockchain systems.

1.3 Internet of Things

Comparing with blockchain, IoT relatively is a type of mature paradigm that combines a group of key components from the perspective of service offerings, namely identification, sensing, computation, communication, services, and semantics [18]. These components are dynamically integrated into one system, where environmental data and users' data are shared/used [19] among five layers, from the perspective of the functionality, including perception, network, middleware, application, and business layers. From the architectural perspective, layers of IoT can be categorized into three layers, which include sensor, transmission, and data processing layers. On-demand services are delivered to users through a network-based channel. Some service offering examples include temperature sensing, traffic monitoring, and route navigation. Specifically, an IoT system assists service providers to achieve offerings by adopting a few metrics, such as availability, reliability, mobility, performance, management, scalability, interoperability, security and privacy. Various metrics are brought by distinct requirements of IoT systems. For example, some focus on data transmission efficiency, while some emphasize security and privacy.

Some typical IoT applications include agricultural networks, unmanned aerial vehicle (UAV) networks, and vehicular ad hoc network (VANET). Because of the rapid integration of things, these applications are merging together to a certain extent due to the broad scope of the connected devices. Additionally, cloud computing and fog computing are flexibly utilized to enhance IoT applications regarding both system functionality and performance.

1.4 Blockchain Applications in Internet of Things

Blockchain applications are under extensive discussions, as technical communities and commercial giants are actively marching towards service-directed systems with specific application requirements. Such systems will enable users to obtain services with more promising features. Originally, blockchain is applied in Bitcoin as a tentative way of conducting financial transactions. The fact that a trusted third party is no longer needed is so fascinating that many blockchain applications as a payment method are proposed, such as Mixcoin, Zerocoin, and Zerocash. Even banks are using it to reconcile banking records happened among them.

Blockchain is mainly used as an immutable ledger in a system where entities lack trust with each other. Given that a centralized storage solution is becoming questionable and inefficient. A blockchain-based solution can improve system transparency, provide record tamper-resistance, and establish trust between entities. This book focuses on five B-IoT systems which are categorized into blockchain-enabled cloud storage services and blockchain-enabled vehicular network services. The blockchain-enabled cloud storage services include blockchain-enabled cloud data preservation services, blockchain-enabled controllable data management. The blockchain-enabled vehicular network services include blockchain-enabled vehicle electricity transaction services, blockchain-enabled carpooling services, and blockchain-enabled ride-hailing services. They are described in detail in the following chapters.

1.5 Security and Privacy in Internet of Things

An IoT system is composed of a large number of devices and communication channels. Both traditional network-related threats and emerging threats exist, thereby IoT systems confront a variety of security and privacy threats, caused by both internal and external adversaries. To establish a reliable and efficient IoT system, it is crucial to recognize specific security and privacy concerns attached to IoT systems, before blockchain technique is involved in the combination.

In most cases, security concerns root in a close attention to IoT data and the vulnerability in underlying computation and communication technologies. Some security concerns include confidentiality, integrity, and authentication. These aspects ensure that data contents are not revealed to adversaries, data packets are not tampered with during transmissions, and unauthorized users cannot access the system, respectively.

Next, privacy concerns are raised along with the enhancement of users' waking awareness for their sensitive information. A privacy consists of two basic parts, which are label (user identity) and the information attached to the label. In an IoT application, e.g., vehicular network, smart grids, and e-healthcare, a user is a main entity who submits data packet containing personal information, such as destinations, smart metering readings, and treatment histories. Privacy can be constructed by knowing both users' identities and the related personal information. Transmitting the data packets without careful sanitization may result in privacy leakage.

Organization of This Book

- Chapter 2 gives an introduction to blockchain and IoT.
- Chapter 3 discusses security issues and privacy concerns in IoT.
- Chapter 4 presents a solution of blockchain-enabled cloud data preservation system.
- Chapter 5 presents a solution of blockchain-enabled controllable data management system.

- Chapter 6 presents a solution of blockchain-enabled vehicle electricity transaction mechanism.
- Chapter 7 presents a solution of blockchain-enabled carpooling services.
- Chapter 8 presents a solution of blockchain-enabled ride-hailing services.
- Chapter 9 discusses several research topics in blockchain-enabled IoT systems.
- Appendix A.1 records how to set up a local Ethereum platform.
- Appendix A.2 is a sample mid-term examination paper.
- Appendix A.3 is a sample final examination paper.

Exercises

1.1 What is blockchain? Refer to Sects. 1.2 and 2.2.

1.2 What is Internet of Things? Refer to Sects. 1.3 and 2.3.

1.3 What are blockchain applications in Internet of Things? Refer to Sect. 1.4.

1.4 What are security issues and privacy concerns in Internet of Things? Refer to Sects. 1.5, 3.2, and 3.3.

1.5 Say Alice is introducing blockchain in Internet of Things, what can she talk about to her friend Bob?

Chapter 2
Blockchain and Internet of Things

2.1 Overview

Bitcoin [1] has attracted extensive attentions from both academia and industry, which initiates a wave of cryptocurrency. It has been widely believed that blockchain [20–24] is the fundamental mechanism for running Bitcoin. Essentially speaking, blockchain technique used in Bitcoin is a public, distributed, and append-only ledger that is maintained and governed by a group of financially motivated miners who do not trust each other. In order to establish a trust environment, transactions (generated by users) are validated by miners and are recorded in a chain of blocks. Blocks are periodically created based on the consensus algorithms appended to the chain. This application formulates a few key characteristics of blockchain, such as distributed ledger storage, transparent operation, and tamper-resistant record.

As mentioned above, the concept of blockchain emphasizes a number of crucial features, such as decentralization, persistency, auditability, and anonymity. These features highlight that blockchain is running upon the decentralized and open context, e.g., peer-to-peer (P2P) network. Blockchain makes use of cryptographic hashes and signatures in order to guarantee both persistency and auditability. Since users can interact with blockchain by multiple addresses (i.e., blockchain identities), real identities/activities can remain anonymous to a certain level.

With the introduction of smart contract, blockchain has been applied or is being applied in various applications transcending its original design purposes. A few representative applications include vehicular networks [25, 26], food supply chain [27], e-health [28], commercial business [29, 30], and industry [31]. The capacity of security and privacy in blockchain also is being enhanced by integrating with security and privacy techniques, e.g., privacy protection [32–34], secure searchable encryption [35, 36], secure multiparty computation [37], access control [38–40], and identity management [41] as depicted in Fig. 2.1.

Moreover, Internet of Things (IoT) [42–49] is enabled by the rapid advancements in multiple technical dimensions, such as sensors, RFID (radio frequency

© Springer Nature Switzerland AG 2019
L. Zhu et al., *Blockchain Technology in Internet of Things*,
https://doi.org/10.1007/978-3-030-21766-2_2

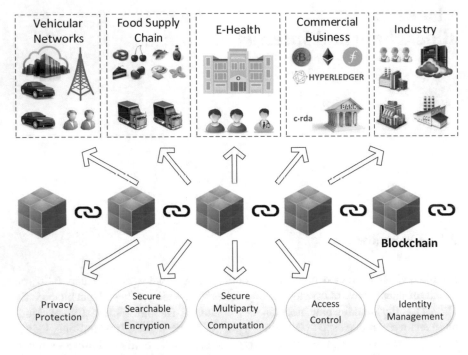

Fig. 2.1 Overview of blockchain

identification), smartphones, computation and communication technologies, and network protocols [50]. Devices and equipment collaborate with one another in the connected environment, which send the collected data (e.g., context or human activities) to a control center or a database. Typical IoT applications include smart grids [51], wireless sensor networks, vehicular networks, smart home, e-healthcare, cloud storage services, agricultural networks, and industrial networks, as illustrated in Fig. 2.2. IoT not only bridges up the physical world and information world, but also enables multi-dimensional data reaching for optimizing decision-making.

This chapter introduces key issues in blockchain and IoT before we dive into detailed discussions. Specifically, this chapter describes technical dimensions of blockchain, technical dimensions of IoT, and typical blockchain applications in IoT.

2.2 Technical Dimensions of Blockchain

2.2.1 Blockchain

We first introduce blockchain covering its model, user, and miner, transaction, incentive, block, chain and height, fork, classification, and smart contract.

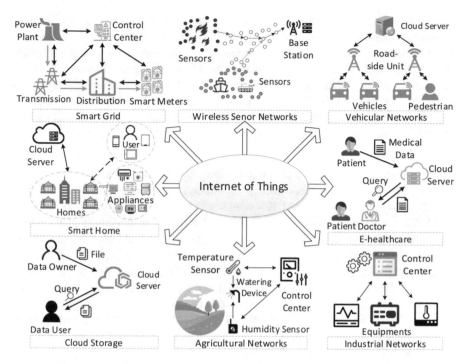

Fig. 2.2 Overview of IoT

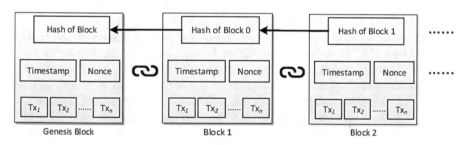

Fig. 2.3 An example of blockchain

Mode In this book, we define blockchain as a distributed and public ledger formed by a chain of blocks in which transactions (Tx) between users, activities, or other key data are recorded. As shown in Fig. 2.3, the initial block in the whole chain is called the genesis block. Hash pointers are used to connect preceding blocks and succeeding blocks. In public blockchain system, miners maintain blocks and ensure the chain to keep growing as long as mining operations are in progress.

User and Miner There are two basic roles in public blockchain: users and miners. A "user" is a person or an organization, which conducts financial transactions

[52] with each other. Users transform their financial transactions into blockchain-designated transactions and upload them to blockchain via sending/broadcasting them to miners. A "miner" is a machine equipped with a blockchain software whose main task is maintaining blockchain. Miners validate all transactions on blockchain and compete to become a winning miner in each time period, relying on the consensus mechanism. A winning miner packs a certain number of transactions to create a block and appends it to the blockchain. Specifically, miners have to invest their own resources (e.g., computational power, electricity, storage) [53] to mining operations.

Transaction   Transaction is a record describing that one party transfers currency to another. For example, Bitcoin currency [1], Ether, is denoted by ETH. The smallest subdenomination is satoshi and 108 satoshis is one "Bitcoin" that is denoted by BTC. A transaction [22, 54] includes inputs and outputs. The input refers to previous transactions through transaction hashes and the index of the transaction output. Each output has an integer that indicates a quantity of BTCs. Each output has a scriptPubKey stating the conditions under which that transaction output can be redeemed. Each output stands by (in a manner of the unspent transaction output (UTXO)) until a later input spends it. Each transaction is hashed by using the SHA-256 algorithm; the resulting hash value is a unique transaction ID. Each transaction is signed by its creator by using the Elliptic Curve Digital Signature Algorithm (ECDSA) and a private key of the signer. A "Locktime" is an unsigned 4-byte integer indicating the earliest time point a transaction can be put to blockchain. Transaction fees can be (not compulsorily) set in each transaction in order to increase the possibility of being selected/packed by winning miners.

Another example [55, 56], Ethereum currency (Ether) is denoted by ETH. The smallest unit of Ether is Wei and one Ether is equal to 10^{18} Wei. A transaction T consists of nonce, gasPrice, gasLimit, to, value, and v, r, s. A nonce Tn is number of transactions sent by the sender; a gasPrice Tp is the amount of Wei to be paid per unit of gas for all computational costs raised from executing the transaction; a gasLimit Tg is the number of gas that is to be used in transaction execution; a "to" Tt is an address of length 160-bit; a value Tv is the amount of Wei to be transferred; v Tw, r Tr, and s Ts are the signatures of the transaction.

Incentive   In the context of digital currency, most miners participate in mining activities for financial purposes. For instance, there are two major sources of incentives in Bitcoin. First, Bitcoin creates a block approximately every 10 min and the winning miner earns BTC(s) as a reward. The reward starts with 50 in 2009 and halves every 4 years on average. The other incentive derives from earning the transaction fee, which is the value difference between a transaction's input and its output.

From an adversarial perspective, when an adversary successfully maneuvers a huge amount of computational power, a 51% attack will be launched to maliciously manipulate the Bitcoin network in the case of the lower-level attack cost. In Bitcoin, an attacker generally Otherwise, adversaries will temporarily follow the rule.

Block Typically, a block consists of the "Block Header" and "Block Body."

A *Block Header* [22] includes: (1) Block version: states a set of block validation rules; (2) Block hash: a 256-bit hash value of the previous block header; (3) Merkle root: a hash value of all the transactions in the block; (4) Timestamp: instant timestamp as seconds; (5) nBits: a hashing target indicating the difficulty of mining, and (6) Nonce: a 4-byte number starts with 0 and increases for every hash calculation.

A *Block Body* includes a transaction counter and transactions. The transaction counter indicates the number of transactions that a block can contain and it depends on block size as well as size of a transaction. All the transactions are validated by ECDSA.

Chain and Height Chain is a virtual string which links a growing set of blocks with cryptographic hashes. As long as blockchain is up and running, the chain will keep growing with new blocks appended in the end. Blocks in the chain are usually addressed by their block height that is a sequence of numbers starting with 0 for the first block, i.e., genesis block.

Fork When two or more miners concurrently create a block, these blocks will have the same block height. This produces a "fork" in blockchain. Forking can be classified into normal occasional forking and rare extended forking [57] as depicted in Fig. 2.4.

Classification Basically, blockchain is categorized into three deployment settings, which are public blockchain, consortium blockchain, and private blockchain. Public blockchain is a public and fully distributed ledger, which does not set up a

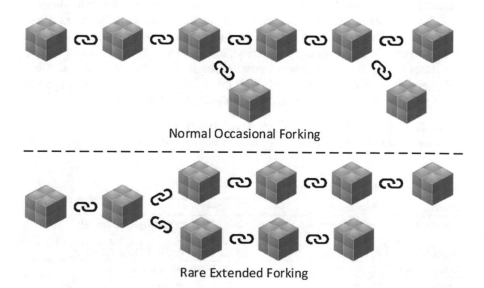

Normal Occasional Forking

Rare Extended Forking

Fig. 2.4 Two types of forks in blockchain

standard for miners to join and its consensus mechanism covers all miners. It is nearly impossible to control the public blockchain, while its efficiency is sacrificed somehow in turn. Consortium blockchain is a partially centralized blockchain for the fact that only a group of miners controls the mining process. It has a certain invitation rule to select miners to participate which leads to a small scale of the consensus mechanism among invited miners. The reading permission can be public or restricted, the efficiency is quite high, but it is possible that a consortium blockchain is tampered. Private blockchain is a centralized blockchain because only one miner is in charge of creating new blocks. The consensus mechanism is carried out by one party. Its reading permission is the same as the consortium blockchain, the efficiency is also high, while it is even more likely to be manipulated and it faces single point of failure. Other classification methods include application (basic blockchain and specific blockchain), independence (main blockchain and side blockchain), and hierarchy (mother blockchain and son blockchain) as listed in Table 2.1.

2.2.1.1 Smart Contract

What is Smart Contract? The concept of smart contract derives from the work in 1994 [65] that used a set of rules to formalize a relationship. The main idea of smart contract is that various types of contractual clauses (e.g., property ownership, bonding) can be embedded in software and hardware and software while ensuring that breaching the contract expensive for malicious breachers.

In blockchain [33], smart contracts are programmable electronic scripts that are typically deployed on the blockchain. Each smart contract has a unique address. A smart contract is triggered by addressing a transaction to it. Then the smart contract self-executes on every node in the blockchain network based on specified codes and data in the triggering transaction, independently and automatically [66]. In another words, the smart contract program logic lies within a "block" that is a software-generated container. The "block" bonds messages pertinent to a specific smart contract. The messages act as inputs or outputs of the smart contract.

What Can Smart Contract Do? A smart contract based blockchain supports multi-step processes between parties who do not trust each other. The transacting

Table 2.1 Blockchain classification

Property	Classification	Example
Openness	Public; consortium; private	BTC, ETH; Hyperledger Fabric[29]; multichain[58]
Application	Basic; specific	ETH, EOS [59]; BTM [60], GXS [61]
Independence	Main; side	BTC, ETH; Mixin network [62]
Hierarchy	Mother; son	NULS [63]; Mixin network; Inchain [64]

parties can review the contract code and predict its outcomes before interacting with the smart contract; (1) have certainty of the contract execution because the code is made public on the network; (2) have verifiability over the outcome because they are signed.

Smart contracts can be applied in many scenarios, such as digital identity, records, security, trade finance, derivatives, financial data recording, mortgages, land title recording, food supply chain, auto insurance, and clinical trials. For instance, smart contracts can offer visibility for a food supply chain at every step. IoT devices write to a smart contract as food moves from its source to a supermarket, providing real-time status of the entire supply chain.

A Simple Example   Smart contracts are mainly adopted in the sense of general-purpose computation. Assume that a blockchain is maintained by John, Mary, and Kerry where digital coins of type A and B are being transferred. John deploys a smart contract that defines: (1) a "deposit" function to deposit 5 units of X into the smart contract; (2) a "trade" function to return 1 unit of A for every 10 units of B it receives; (3) a "withdraw" function to withdraw all the coins in the smart contract. The "deposit" and "withdraw" functions can be called by John or by any user on the network according to application requirements.

Now John sends a transaction to the smart contract's address which moves 5 units of A to the smart contract by call its "deposit" function. Next, Mary sends a transaction to the smart contract's address which moves 20 units of B to the smart contract by call its "trade" function, and obtains 2 units of B in return. The above two transactions are recorded on the blockchain. Finally, John sends a transaction with a signature to the smart contract's "withdraw" function. After the signature is verified, the smart contract sends all deposits (3 units of A and 20 units of B) to John.

Smart Contract in Ethereum   Ethereum is the first Turing-complete decentralized system and it is built for creating smart contracts. Ethereum replaces the restrictive language in Bitcoin with a language enabling developers to write personalized programs or autonomous agents. There are two types of nodes in Ethereum: Ethereum virtual machine (EVM) and mining node. EVMs' purpose is to execute the codes in smart contracts and miners' job is to write transactions to the Ethereum blockchain. With Ethereum's assistance, companies and researchers are building multiple smart contract applications based on Ethereum, such as supply chain [67], crowdfunding [68], and security and derivatives trading [69].

Security Issues   Since smart contracts in blockchain are visible to all users, this results in bugs and vulnerabilities that may not be fixed in time. SpankChain, an adult entertainment platform built on Ethereum, has suffered a breach that costs nearly $40,000 [70]. The team said, the attack was raised from a "reentrancy" bug that was similar to the one in the hack of Distributed Autonomous Organization (DAO) project in 2016. The attacker built a malicious contract disguising as an ERC20 token, where the "transfer" function is called back into the payment channel contract for several times, stealing some Ethers every time.

Thirty-four thousand two hundred vulnerable contracts have been found through a scan of nearly one million Ethereum smart contracts. These contracts can be utilized to steal ETH, and freeze or even delete assets in smart contracts [71]. Some typical vulnerabilities include transaction ordering dependence, timestamp dependence, reentrancy, forcing Ether to a contract, and DoS with block gas limit. When writing Ethereum smart contracts, the most crucial thing to do is provide security. Otherwise, we will suffer from huge loss.

2.2.2 Consensus Mechanisms

In blockchain, the consensus mechanism is a method of reaching a consensus among untrustworthy nodes, which is a variation of the byzantine generals problem [72]. In this problem, several Byzantine generals who separately command a group of army are surrounding a city and preparing to attack it. The attack would succeed only if all of the generals attack the city together. So they have to agree upon when to attack. But in this trustless environment, there could be traitors among generals who deliver different messages to different generals. Therefore, how to reach a consensus among generals is a challenging task.

For blockchain, it is also challenging to reach a consensus among miners, because blockchain network is distributed and no centralized role exists to manage the network. Miners do not trust each other. Next, we present four typical consensus mechanisms in blockchain.

Proof-of-Work Proof-of-Work (PoW) [1] requires miners to ceaselessly calculate a cryptographic hash function SHA-256 using a block header and different nonce until a resulting value is smaller than a given value, i.e., target threshold or just difficulty. It is like seeking a solution to a hash puzzle. Here, a good hash function maps any input into a random number. If the input is changed even in a bit, the produced hash result will be completely different number, thus making hash function unpredictable. After a miner obtains a solution to the hash puzzle, all other miners have to confirm its correctness. If it is valid, transactions used for the hash calculation are approved to be added to blockchain.

The meaning of PoW is that miners have to physically do some computations and compute to be the first one who finds a provable solution. The combination of solution seeking with award winning is called mining, as illustrated in Fig. 2.5

Unfortunately, miners have to contribute computation resources during PoW mining in order to win. This leads to a great investment of computation hardware from both organizations and individuals, let alone huge electricity consumption. For this reason, some PoW based protocols with additional functions have been proposed. For example, Primecoin [73] allows miners to search special prime numbers during mining which is useful to prime number research. Additionally, if an adversary holds a 51% of the whole computation power, then it can control PoW process so as to control blockchain with a high probability.

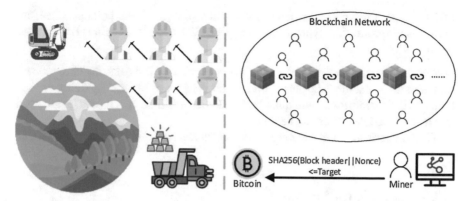

Fig. 2.5 An analogy between actual mining and PoS mining

Proof-of-Stake Proof-of-Stake (PoS) [74, 75] only needs miners to run an algorithm that randomly chooses one miner to be the winning miner proportionally to miners' stakes. PoS is a natural alternative mechanism to PoW.

Kiayias et al. [75] proposed a blockchain protocol based on PoS and they assume that miners can arbitrarily create accounts and make payments with their stakes changing over time. The protocol was presented with four phases successively improving the adversary model. For instance, in the first static phase, i.e., static stake, an initial stake distribution was embedded in the first block including miners' public keys $\{pk_i\}$ and stakes $\{s_i\}$. An ideal function $\mathscr{F}$ was available to all miners and took as input stake set $\mathbb{S}$, auxiliary information ρ, and time slot sl, and output a winning miner M_i with probability

$$p_i = \frac{s_i}{\sum_{k=1}^{n} s_k}. \tag{2.1}$$

The above selection process can be achieved in a straightforward manner. For example, let $\hat{p}_i = s_i / \sum_{j=i}^{n} s_j (1 \leq i \leq n-1)$ and $\hat{p}_n = 1$. The process starts with flipping a $\hat{p}_i$-biased coin. If the result is 1, then an M_i is selected. Or it continues to flip the next coin.

Compared with PoW, PoS is more energy efficient and is more effective in reaching a consensus. However, it can lead to situation where only a few miners with higher stakes keep adding new blocks to blockchain. To combine the benefits of PoS and PoW, Ethereum is preparing to transfer from PoW to Casper [76], where miners have to pay some deposits to verify a block. If this block is added to blockchain later, then the verifiers of the block will be rewarded proportionally to their deposits.

PBFT Practical Byzantine fault tolerance is a state machine replication algorithm [77, 78] that tolerates Byzantine faults. It works in asynchronous systems and provides liveness and safety if at most $\lfloor \frac{n-1}{3} \rfloor$ of replicas are faulty at the same time. It requires nodes to co-maintain a state and act consistently. To achieve this goal,

PBFT designs three basic protocols: consistency protocol, checkpoint protocol, and view-change protocol. A typical example of PBFT-based blockchain is Hyperledger fabric [29].

DPOS Delegated proof-of-stake (DPOS) is similar to PoS, but the difference is that miners elect some delegates to create and verify new blocks. By doing so, new blocks will be confirmed more quickly. Block size and time epochs can be adjusted, and misbehaved delegates can be voted out.

The comparison of the above four consensus mechanisms is listed in Table 2.2.

2.2.3   Key Characteristics

Blockchain has four key characteristics [22]:

- **Decentralization**. In traditional centralized online business systems, transactions have to be verified by a centralized server, such as a bank. This approach inevitably leads to performance bottlenecks of communication overhead for the whole system and computational cost for the centralized server. A blockchain network is built on P2P network and transactions can be conducted between any two entities without the participation of the centralized server. In this way, blockchain greatly slashes the cost of the server and mitigates the performance bottlenecks.
- **Persistency**. Since all broadcasted transactions in blockchain are validated and then recorded in blocks, it is difficult to falsify them. Moreover, blocks are verified by other miners, it is also nearly impossible to tamper with them.
- **Auditability**. Given that all transactions in blockchain are signed by its sender and then recorded in a block with a timestamp, users can trace and validate information in the transactions. Additionally, each transaction is linked to a previous transaction iteratively which enables source traceability and data transparency.
- **Anonymity**. Each user holds a self-generated address to interact with blockchain every time. Additionally, the user can generate a set of addresses in advance just to keep her/his identity anonymous. Since there is no longer a centralized role to

Table 2.2 Blockchain consensus mechanism

Type	Miner management	Electricity cost	Adversary tolerance	Example
PoW	Open	High	<51% computational power	BTC, ETH
PoS	Open	Some	<51% stakes	Casper
PBFT	Permissioned	A few	<33.3% replicas	Hyperledger Fabric
DPOS	Open	Some	<51% delegators	Bitshares [79]

keep a list of users' real identities, the risk of identity exposure is highly reduced. By doing so, blockchain protects user privacy to a certain degree.

2.2.4 Applications of Blockchain

There are various blockchain applications in the literature. These applications include vehicular networks [25, 26], food supply chain [27], e-health [28], commercial business [29, 30], and industry [31] as shown in Fig. 2.1. Most of them leverage decentralization and persistency of blockchain and apply it in their scenarios with specific features and requirements. This subsection only gives a brief introduction to blockchain applications in food supply chain and commercial business. A detailed description of blockchain applications in IoT is given in Chaps. 4–8.

Food safety is now a worldwide problem [80] given that a bunch of grave food safety incidents happened, such as "Sudan red" [81] and "horsemeat scandal" [82]. There is an urgent need to build a food supply chain traceability system. Tian [83] proposed a blockchain-based supply chain traceability system (SCTS), as depicted in Fig. 2.6. SCTS makes use of RFID, sensors, and GPS signals to collect information and utilize BigchainDB [84] to store data in food supply chains, where BigchainDB takes a distributed database and gives it blockchain characteristics. The system model of SCTS consists of the producer, processor [85], authorized organization, distributor, retailer, consumer, and user. Each legal entity can review, insert, and update the information on BigchainDB. Each food product is appended with a RFID tag as a digital identifier. Each user also has a digital profile, which includes identity, location, certifications, and links to food products. SCTS is

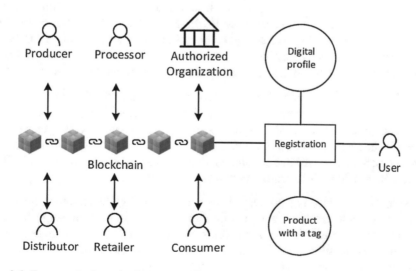

Fig. 2.6 Framework of supply chain traceability system

designed with a group of rules that regulate how users interact with the system, and how data is shared. After rules are written in BigchainDB, they cannot be changed privately unless being broadcasted to and verified by miners.

Hyperledger Fabric [29] is an open-source and distributed operating system for permissioned blockchains and it is used in more than 400 prototypes across different application scenarios. Hyperledger Fabric presents a novel blockchain architecture that provides benefits such as resiliency, flexibility, scalability, and confidentiality. It is designed with a general purpose for permissioned blockchain which supports the implementation of distributed applications created in standard programming languages. By doing so, Hyperledger Fabric becomes the first distributed operating system for permissioned blockchains. It also securely records execution histories in an append-only ledger and does not have cryptocurrency.

The architecture of Hyperledger Fabric is an execute-order-validate paradigm and it partitions transaction flow into executing and checking a transaction; ordering via a consensus protocol; and transaction validation according to trust assumptions. A distributed application of Hyperledger Fabric is composed of (1) a smart contract, named chaincode, which executes application logic in form of programming codes and (2) an endorsement policy that servers as a static library for transaction validation. A Hyperledger Fabric blockchain network consists of a set of nodes that can be clients, peers, and ordering service nodes.

2.3 Key Issues in Internet of Things

2.3.1 Fundamental Concepts in Internet of Things

We first introduce IoT regarding its fundamental concepts in IoT, as illustrated in Fig. 2.6. Identification and sensing are the basic functions of IoT, which transmit device information and sensed data to computation components through communication protocols. Finally, services are provided based on the information and data, from which semantics will be extracted to utilize services and business models.

Identification Identification [45] is important for naming IoT devices and providing services to them. Some identification methods include ubiquitous codes and electronic product codes [86]. Addressing IoT devices is vital to distinguish device ID and its address. Device ID is a device name and it is not globally unique, while its address is an address within a communications network that can uniquely identify a device. Here, addressing methods include IPv4 and IPv6.

Sensing Sensing [45] is a concept of collecting data (e.g., temperature, humidity, pressure, carbon dioxide, motion, vibration, acceleration) from some targeted objects within a certain range of environment and uploading it to a control center or a database. Sensing is carried out by sensors with limited power. Sensors can be pre-deployed in a target area or wore on human bodies.

Belkin WeMo [87] enables people to remotely control home appliances based on movements with a sensor plugging into an outlet. The sensor can capture movements up to ten feet away and send a command to a WeMo switch wirelessly to turn the appliance off or on.

Computation After data is collected, it will be analyzed by IoT devices or managers to take corresponding actions based on specific demands. Computation is conducted by processing units (e.g., micro-controllers, micro-processors, field programmable gate array (FPGA)), software (e.g., TinyOS [88], Riot OS [89]), and hardware [90] (e.g., Intel Galileo, Raspberry PI).

Except for the 1/0 decision-making as in a switch control, some typical computation functions are summary, average, median, maximum, minimum, and variance. To complete these functions, an aggregator node [91–95] is introduced. For instance, in-network aggregation of summary is proceeded by letting each intermediate node send a single message containing a sum of the sensor readings of nodes in previous transmissions. In this way, not only a summary of sensor readings can be acquired in the end, but also saves energy costs for each node, which is important for energy-constrained sensors.

Communication IoT communication technologies [45] connect various devices together to provide services. Generally speaking, IoT devices are energy-constrained [96, 97] and running in a noisy communication environment. Communication protocols for IoT should adapt to these unique features. Some typical communication protocols are radio frequency identification (RFID), near field communication (NFC), Wi-Fi, Bluetooth, IEEE 802.15.4, and long-term evolution (LTE).

Services IoT services are classified into four classes [98, 99]: identity-related services, information aggregation services, collaborative-aware services, and ubiquitous services.

- **Identity-related services**: provide the most fundamental services for other services since identification is a priority when linking an object from the real world to an entity in the virtual world.
- **Information aggregation services**: gather and process sensed data.
- **Collaborative-aware services**: use the sensed to make decisions.
- **Ubiquitous services** offer the third services when they are in need.

Semantics IoT semantic is an ability to acquire knowledge via various devices to deliver services. It includes knowledge discovery, knowledge utilization, and information model. Additionally, it includes processing data to make right decisions in order to offer exact services. Semantic is the core of IoT by sending demands to the right resource. A classic semantic web technologies are resource description framework (Fig. 2.7).

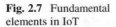
Fig. 2.7 Fundamental
elements in IoT

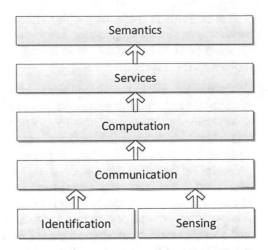

2.3.2 Architecture of Internet of Things

Given that IoT hash connected billions of devices and utilized many computation
and communication technologies among them, a clearly layered architecture would
be nice for understanding IoT from a high level. There is a five 5-layer model for
IoT [100], namely perception layer, network layer, middleware layer, application
layer, and business layer, as illustrated in Fig. 2.8:

- **Perception Layer**: is a device/object layer. It is composed of physical devices.
  This layer handles device identification and data collection. The collected data is
  sent to the next layer, i.e., network layer, for further data processing.
- **Network Layer**: is also known as "transmission layer." It securely sends data
  from devices to a data processing system in a wired or wireless way.
- **Middleware Layer**: consists of a variety of devices. IoT devices serve for
  various services and they only communicate with the ones of the same service
  description. This layer is in charge of managing services and a database. It
  receives data from network layer and records them in the database. Then it
  processes the data to make corresponding decisions.
- **Application Layer**: offers the management of applications based on processed
  data in the middleware layer and delivers services requested by customers.
- **Business Layer**: manages the overall IoT system including applications and
  services. It helps companies build business models, obtain analysis results, and
  determine future strategies.

Fig. 2.8 A 5-layer model of
IoT

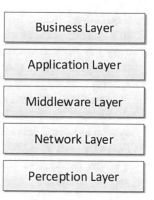

Business Layer

Application Layer

Middleware Layer

Network Layer

Perception Layer

2.3.3 Evaluation Metrics of Internet of Things

When evaluating IoT systems, we can resort to some common metrics [45, 101], which covers a few aspects, such as availability, reliability, mobility, performance [102, 103], management, scalability, interoperability, security and privacy. Addressing these issues will assist IoT service providers and engineers in realizing services more pertinently.

- **Availability**: Availability means service delivery for customers anywhere and anytime at both software and hardware levels. Software availability is an ability to deliver services for customers anywhere simultaneously. Hardware availability is the existence of devices running software to deliver services anytime.
- **Reliability**: Reliability is an operation state of an IoT system. High reliability stands for a high success rate of service delivery which is tightly close to availability. Compared with availability, reliability is given more rigorous requirements especially when the IoT application is related to emergency response.
- **Mobility**: Mobility refers to service continuity given the fact that many IoT services are provided to mobile devices facing service interruptions. Connecting these devices moving at a high speed and continuously delivering corresponding services is crucial to IoT services.
- **Performance**: Performance depends on software, hardware, and underlying computation and communication technologies. Many specific aspects can be adopted to assess the performance, such as computational cost, communication overhead, response delay, and storage cost.
- **Management**: Numerous devices and applications are deployed and provided in IoT systems which brings IoT managers some pressing issues to handle, such as configuration, coordination, performance, malfunction, and security. Managing IoT devices efficiently will lead to an effective growth of IoT systems, while efficient management schemes are much preferred.

- **Scalability**: Scalability is an ability to introduce new devices and services to IoT systems without adversely impacting previous service quality. This is not an easy job since new devices and services may be not compatible with existing ones regarding data format, communication protocols, etc.
- **Interoperability**: Interoperability addresses the issue that many heterogeneous devices from different manufacturers should be harmonized by IoT application developers. Interoperability is significant for building a comprehensive IoT system to ensure service delivery.
- **Security and Privacy**: Security and privacy presents a vital challenge for IoT systems because of the existence of various attacks and a lack of common security standard. In general, security mainly refers to system security and data security. Privacy includes identity, location, trajectory, report, and query. The next chapter will give a full description on security issues and privacy concerns in IoT.

2.3.4   Function Enhancement

Cloud computing (CC) CC [104–108] is a dynamic combination of configurable computer resources and multiple services. It has been developing over the past decade and generated various capabilities and strengths. CC [109, 110] has a high-level cloud server with powerful computation power and a massive storage space as well as devices (users). The architecture of CC is shown in Fig. 2.9.

Classic cloud platforms include Google Cloud, Amazon Cloud, and Alibaba Cloud. For instance, Google Cloud Platform [111] offers advanced analytics and machine learning services to extract valuable information from uploaded data.

Fig. 2.9 Architecture of cloud computing

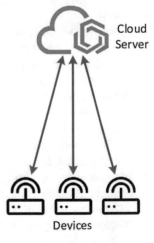

Amazon Web Services [112] provide a set of cloud storage services for data archiving. People have an option to choose Amazon Glacier for non-expensive data storage, and Amazon Simple Storage Service for data storage. Thus, CC can be used in IoT applications to perform computation tasks and alleviate computation burdens of devices.

However, adopting CC in IoT applications is challenging because of the following five problems:

- **Delay**: Since data are uploaded to a cloud server that processes the data for low-level devices, it increases the response delay. How to provide real-time services for users pose a challenge.
- **Standardization**: If two or more cloud servers are interoperated, how to achieve standardization of data format, communication protocol, and computation synergy is also a challenge.
- **Balancing**: The CC environment and IoT system requirements are naturally discordant due to infrastructure differences which is also another challenge.
- **Security**: Security mechanisms for CC and IoT are different and they have to be carefully addressed before a secure combination can be achieved. Especially when the cloud server in CC is not always trusted and it may violate users' privacy.
- **Reliability**: Providing a reliable CC-IoT system is not easy for the reason that have different devices and resources.

Fog Computing (FC) In the year of 2012, the concept of fog computing is proposed [113]. FC is a distributed computing paradigm that utilizes devices at network edge to perform some of the computation, storage, communication for the cloud server in a local manner. It consists of a similar high-level cloud server, devices (users), and a fog node. Generally speaking, there can be several fog nodes that are deployed in a wide service area. The architecture of FC is shown in Fig. 2.10.

FC acts as a bridge between the cloud server and devices. It extends CC's capabilities to the network edge and processes sensed data locally using a fog node similar to a mini-cloud server. Since fog nodes have proximity to users compared to the cloud server, FC has the ability to provide services of quick delivery and low delay, increasing the overall performance of IoT applications.

Specifically, FC has the following features [114, 115]:

- **Geographic Distribution**: The fog nodes are located in places, such as roadside, near cellular base stations, at a point-of-interest. They are deployed in a distributed manner to guarantee that fog nodes will collect real-time data feeds from users.
- **Location Awareness**: Locations of fog nodes are known to the network manager and cloud server. FC aims at providing location-based services for users at certain areas via fog node. Hence, FC can be aware of users' locations according to locations of fog nodes.
- **Low Latency**: Given some capabilities of computation, storage, and communication, fog nodes are able to make decisions from local data without a cloud server.

Fig. 2.10 Architecture of fog
computing

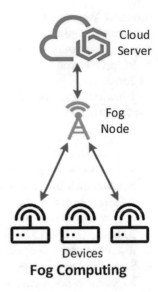

Moreover, the response latency is further reduced since fog nodes are close to
end users.

- **Management Support**: FC can support large-scale IoT applications which
  previously brought heavy management overhead of device and application to
  the cloud server. Considered that fog nodes can autonomously manage a set of
  devices, the centralized cloud server is relieved from such a burden.
- **Partial Decentralization**: FC is a partial decentralized architecture in the sense
  that the cloud server can be removed from managing devices and services. The
  fog nodes themselves can cooperate with each other to provide services to users.

Fog-assisted IoT services [115] include real-time services (e.g., smart traffic
lights, healthcare and activity tracking, vehicular navigation), transient storage
(e.g., edge content caching, shopping cart management, software and credential
updating), data dissemination (e.g., energy consumption collection, local content
distribution, malware defense), and decentralized computation (e.g., computation
offloading, aided computation, big data analysis).

FC cannot be considered secure since it inherits many security threats from CC.
One common assumption is that fog nodes are honest-but-curious. They are built
and deployed by commercial fog manufacturers to provide services to earn profits.
On the one hand, fog nodes strictly follow the prescribed protocols. On the other
hand, they may probe into data contents. Moreover, employees of FC companies
may obtain users' personal information. Additionally, fog nodes may become major
targets of hackers who misuse the leaked data.

2.3.5 Applications of Internet of Things

IoT applications include smart grid, wireless senor networks, vehicular networks, smart home, e-healthcare, cloud storage services, agricultural networks, industrial networks as illustrated in Fig. 2.2.

- **Smart Grid**: Millions of smart meters [116] are deployed in the world to collect real-time electricity consumption [117] to power companies in smart grids [118]. The power companies can record and estimate electricity usage and demand according to top gathered smart meter readings. For customers, smart grids can provide them with reliable power services and flexible pricing policies.
- **Wireless Sensor Network**: A group of sensors for a local wireless sensor network and periodically senses environmental data to report to a base station. A wireless sensor network is usually deployed in an unmanned area and sometimes severe environments. Sensors are always energy-constrained and lightweight computation and communication protocols should be used to prolong network lifetime.
- **Vehicular Networks**: A vehicular network [119] consists of user (vehicles, pedestrians), road-side unit (RSU), and cloud server. Running vehicles and walking pedestrians are equipped with on-board units (OBUs) and mobile devices, respectively. They can interact with the network to share information, such as road congestion, navigation, parking spot finding [120].
- **Smart Home**: Modern home appliance can establish a smart home network with a cloud server and a controller at user end. The user is able to send control commands to her/his appliances through the cloud server. In turn, appliances with sensors can report working status back to the user if some preset conditions are met, e.g., sunlight shoots onto the appliance.
- **E-healthcare**: Hospitals nowadays have their own database to store patients' record, including their identity, type of illness, prescription, assigned doctor, etc. With such a database, doctors can access patients' treatment histories more conveniently in order to make proper decisions in future treatment.
- **Cloud Storage Services**: With the development of CC, more and more users are willing to store their files on a cloud server, so they will not have to carry a storage device anywhere they go if they want to access her/his files. Besides, a cloud server also provides some computation functions for users to enjoy, such as search, sort, and deduplication.
- **Agricultural Networks**: Modern agricultural networks enable farmers to remotely monitor the status of plants based on readings reported by sensors. They can also execute agricultural operations automatically by sending commands to on-site devices.
- **Industrial Networks**: Industrial networks are composed of control center and equipment. They can communicate with each other to share the latest status of equipment and keep the whole system running correctly.

This book focuses on cloud storage services and vehicular networks, and will introduce some state-of-the-art blockchain applications in these two applications from Chaps. 4 to 8.

2.4 Summary

This chapter introduces the technical dimensions of blockchain and key issues in IoT. Although IoT has matured to a huge and powerful system, blockchain is still developing rapidly. Considered that people in IoT systems are more aware of the importance of features brought by blockchain, there are still many potential opportunities for the combination of blockchain and IoT.

Exercises

2.1 What is blockchain? Refer to Sect. 2.2.

2.2 What are the crucial components of blockchain? Refer to Sect. 2.2.

2.3 Say Alice is a user in the Bitcoin network, and Bob is the miner in the Bitcoin network, what is the difference between Alice and Bob regarding operations in the Bitcoin network?

2.4 Why do miners participate in blockchain mining? Refer to Sect. 2.2.

2.5 Why does forking happen in blockchain mining? Refer to Sect. 2.2.

2.6 What are the classification of blockchains? Refer to Sect. 2.2.

2.7 What is smart contract and what can it do? Refer to Sect. 2.2.

2.8 Please give an example of how a smart contract work. Refer to Sect. 2.2.

2.9 How do smart contract have security flaws? Refer to Sect. 2.2.

2.10 How to reach a consensus in a blockchain network? Please give some examples. Refer to Sect. 2.2.3.

2.11 What are the key features of blockchain? Refer to Sect. 2.2.3.

2.12 Please give some blockchain applications. Refer to Sect. 2.2.4.

2.13 What are the crucial components of IoT? Refer to Sect. 2.3.1.

2.14 What are the five layers of IoT? Refer to Sect. 2.3.2.

2.15 How to evaluate an IoT system? Refer to Sect. 2.3.3.

2.16 Is there any enhancement methods for IoT application? Refer to Sect. 2.3.4.

2.17 Please briefly introduce some IoT applications. Refer to Sect. 2.3.5.

Chapter 3
Security and Privacy Issues in Internet of Things

3.1 Overview

IoT systems of a high level of ubiquity and heterogeneity are confronted with various security and privacy threats [115, 121–125]. In order to guarantee system functionality and reach a complete acceptance of users, it is necessary to specify security issues and privacy concerns in IoT.

The security issues mainly include confidentiality [126, 127], integrity [128], and authentication [129, 130]. Generally speaking, confidentiality ensures that data contents are not revealed to adversaries [131]; integrity [132] guarantees that data packets are not tampered with during transmissions; and authentication [133] prevents unauthorized users from accessing the system. Privacy concerns [104, 134–136] arise from the fact that data packets transmitted from users to IoT infrastructures may contain sensitive information (e.g., identity, location, trajectory, report, query). Since the information is highly related to user privacy, leaking it will expose user to attacks ranging from advertisement spams, to stocking, or even physical injury. Therefore, privacy enforcements must be provided by IoT systems.

Many protection mechanisms have been proposed in different IoT scenarios. These proposed mechanisms aimed at solving designated security problems, since system requirements and security models differ for different applications. This chapter presents main security issues and privacy concerns in IoT systems, as shown in Fig. 3.1. Introducing the two aspects will help researchers understand security and privacy problems in IoT systems and come up with more pertinent protection mechanisms.

© Springer Nature Switzerland AG 2019
L. Zhu et al., *Blockchain Technology in Internet of Things*,
https://doi.org/10.1007/978-3-030-21766-2_3

Fig. 3.1 Main security issues
and privacy concerns in IoT

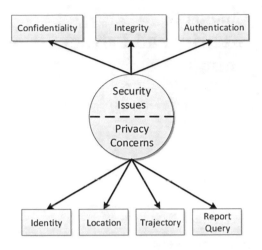

Fig. 3.1 Main security issues
and privacy concerns in IoT

3.2   Security Issues in Internet of Things

Security issues mainly consist of three aspects: confidentiality, integrity, and authentication. In this section, we introduce security issues and corresponding countermeasures in different IoT scenarios.

3.2.1   Confidentiality

Confidentiality, i.e., data confidentiality, means that data contents during transmissions are not leaked to any adversary. Data content in IoT systems usually refers to the plaintext content generated by a user before she/he performs any complicated operations, such as encryptions [137, 138] and perturbations. Generally speaking, confidentiality is protected by encryption if indistinguishability is guaranteed in front of an adversary. In order to protect data contents from the adversary, we firstly model the adversary and its ability. Assume that an adversary $\mathscr{A}$ is a probabilistic polynomial-time (PPT) adversary [139] which can launch four types of attacks:

- **Ciphertext-Only Attack**: This is the lowest level attack which is considered the most common one. In this case, the adversary only observes a communication channel and ciphertexts in it, and tries to obtain underlying plaintexts.
- **Known-Plaintext Attack** (KPA): It refers to a scenario where the adversary can acquire some pairs of plaintext/ciphertext generated under a secret key.
- **Chosen-Plaintext Attack** (CPA): This attack is a little bit more destructive than the first two attacks since the adversary is able to obtain ciphertexts for self-chosen plaintexts.
- **Chosen-Ciphertext Attack** (CCA): In the fourth attack, the adversary can obtain plaintexts for self-chosen ciphertexts.

Here, we introduce the third attack and the corresponding security in details, as an example.

First, a symmetric encryption scheme $\prod$ consists of a message space M, an algorithm Gen, an algorithm Enc, and an algorithm Dec:

- M is the set of valid inputs that are supported by the encryption scheme.
- Gen is a probabilistic algorithm which outputs a key k.
- Enc takes a key k and a message m as input, and outputs a ciphertext c.
- Dec takes a key k and a ciphertext c as input, and outputs a plaintext m.

A symmetric encryption scheme $\prod$ satisfies a requirement: for every key k output by Gen and every message $m \in M$, it holds that

$$\text{Dec}_k(\text{Enc}_k(m)) = m.$$

Second, CPA is modeled by giving an adversary $\mathcal{A}$ access to an encryption oracle $\text{Enc}_k(m)$ that encrypts A's chosen messages under a key s that is unknown to $\mathcal{A}$. When $\mathcal{A}$ queries $\text{Enc}_k(.)$ by giving m, $\text{Enc}_k(.)$ returns $c = \text{Enc}_k(m)$. When Enc is randomized, $\text{Enc}_k(m)$ chooses new randomness in every query.

Third, a CPA indistinguishability experiment $\text{PrivK}_{\mathcal{A},\prod}^{\text{cpa}}$ is constructed as follows:

- A key k is created by $\text{Gen}(1^n)$, where n is a security parameter.
- $\mathcal{A}$ is given 1^n and access to $\text{Enc}_k()$, and generates two messages m_0, m_1 of the same length.
- A bit $B \in \{0, 1\}$ is chosen uniformly. A ciphertext $c = Enc_k(m_B)$ is calculated and sent to $\mathcal{A}$.
- $\mathcal{A}$ continues to query $\text{Enc}_k()$ and outputs a bit B'.
- The output of $\text{PrivK}_{\mathcal{A},\prod}^{\text{cpa}}$ is 1 if $B' = B$, and 0 otherwise. $\mathcal{A}$ succeeds if $B' = B$.

Last, the formal CPA-security is defined as:

Definition 3.1 (CPA-Security) A symmetric encryption scheme $\prod = $ (Gen, Enc, Dec) is CPA-secure, if for all PPT adversaries $\mathcal{A}$, there is a negligible function neg such that

$$\Pr[\text{PrivK}_{\mathcal{A},\prod}^{\text{cpa}}(n) = 1] \leq 1/2 + \text{neg}(n).$$

Existing Work There are numerous works concentrating on confidentiality protection in IoT. We look into some of them and discuss how confidentiality matters in different scenarios.

Wireless sensors [140] have been deployed in applications such as road traffic monitoring, and forest fire tracking. In a typical sensor network, as depicted in Fig. 3.2, thousands of sensors with low communication and computation capabilities are organized into a tree or a cluster before a base station broadcasts a collection task (the task can be embedded into sensors). The sensors periodically collect environmental data and report pre-processed data to the base station for data analysis.

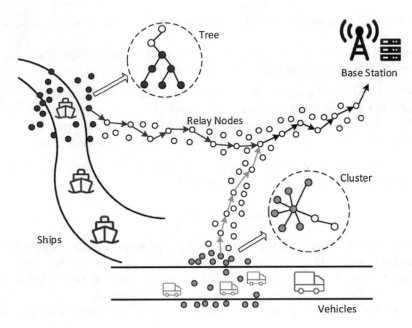

Fig. 3.2 Architecture of a typical sensor network

Confidentiality is crucial in such scenario [141–143] since sensitive information collected will be leaked to adversaries if confidentiality is compromised. Li et al. [140] proposed a provably secure aggregation scheme to preserve confidentiality. Specifically, they perturb environmental data with hash values and require a base station to iteratively compute the right combination of hash value set and then decide which sensors contribute data after receiving a final aggregation result. Since each reading reported by each sensor is added with a hash value and the base station can only acquire the sum of all sensed readings, no entities other than the sensor itself can reveal the true reading. Thus, confidentiality is protected.

3.2.2   Integrity

Integrity, i.e., data integrity refers to the property that data contents during transmissions are not altered by any adversary. Generally speaking, integrity is protected by digital signatures if existential unforgeability is guaranteed in front of a PPT adversary.

A PPT adversary can launch three types of attacks [139] as follows:

- **Random-Message Attack** (RMA): The adversary cannot control messages that are signed and it can only observe signatures generated by honest entities on messages.

- **Known-Message Attack** (KMA): The adversary has a limited control over what messages are signed, which means that the adversary has to specify messages in advance independent of the public key of the signer and following signatures.
- **(Adaptive) Chosen-Message Attack** (CMA): The adversary has a complete control over what messages are signed which means that the adversary can select messages after it observes the signer's public key and previous signatures.

Here, we introduce the second attack and the corresponding security in detail as an example.

First, a signature scheme $\prod$ consists of a message space M, an algorithm Gen, an algorithm Sign, and an algorithm Vrfy:

- M is the set of valid inputs that are supported by the signature scheme.
- Gen is a randomized key-generation algorithm that takes security parameter n as input. It outputs a private (signing) key sk and a public (verification) key pk.
- Sign is a signing algorithm which takes n, sk and a message m as input. It outputs a signature σ.
- Vrfy is a verification algorithm which takes n, pk, m, and σ as input. It outputs 1 if σ is a valid signature on m under pk and 0 otherwise. A message/signature pair (m, σ) is valid if $\text{Vrfy}(m, \sigma) = 1$.

A signature scheme $\prod$ satisfies a requirement: for all n, (sk, pk) generated by $\text{Gen}(n)$, $m \in \{M\}$, and σ generated by $\text{Sign}_{sk}(m)$, it holds that $\text{Vrfy}(m, \sigma) = 1$.

Second, KMA is modeled by giving an adversary $\mathscr{A}$ access to a signing oracle $\text{Sign}_{sk}(m)$ that encrypts A's l chosen messages $m_1, \ldots, m_l$ under a private key sk that is unknown to $\mathscr{A}$. When $\mathscr{A}$ queries $\text{Sign}_{sk}(.)$ by giving $m_1, \ldots, m_l$, $\text{Sign}_{sk}(.)$ returns signatures $\sigma_1 = \text{Sign}_{sk}(m_1), \ldots, \sigma_l = \text{Sign}_{sk}(m_j)$.

Third, a KPA existential unforgeability experiment $\text{PubK}^{kpa}_{\mathscr{A}, \prod}$ is constructed as follows:

- A pair of keys (sk, pk) is generated by $\text{Gen}(n)$.
- $\mathscr{A}$ outputs l messages $m_1, \ldots, m_l$.
- l signatures $\sigma_1 = \text{Sign}_{sk}(m_1), \ldots, \sigma_l = \text{Sign}_{sk}(m_j)$ are calculated.
- $\mathscr{A}$ is given pk and $\sigma_1, \ldots, \sigma_l$, and outputs (m, σ).
- $\mathscr{A}$ succeeds if $\text{Vrfy}_{pk}(m, \sigma) = 1$ and $m \{m_1, \ldots, m_l\}$.

Last, the formal KMA-security is defined as:

Definition 3.2 (KMA-Security) A signature scheme (Gen, Sign, Vrfy) is existentially unforgeable under a KMA if for all polynomials l and all PPT adversaries $\mathscr{A}$, $\mathscr{A}$ has a negligible success probability in the above experiment.

Existing Work Smart grid (SG) [144] is considered the next generation of power grid [118, 145, 146]. It combines traditional grid with information technologies to enable bidirectional transmission. Meanwhile, it aims to provide reliable, efficient, sustainable, customized, and secure power generation. Smart meter (SM) is an important component in smart grid and is a bidirectional communication device installed in consumer' house. It periodically collects real-time electricity consump-

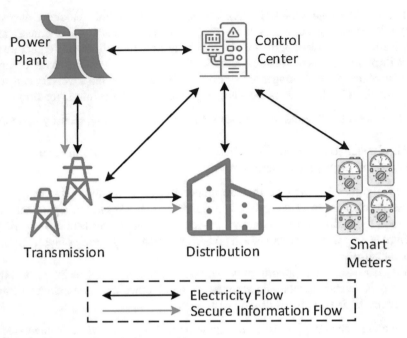

Fig. 3.3 Architecture of a typical smart grid

tion and reports meter readings to an operation center via a reliable communication network depicted in Fig. 3.3.

Meter readings are crucial to monitor information about grid operations and status. If they are tampered with or forged by an adversary, it will lead to supply estimation chaos for grid and financial losses for consumers. Therefore, integrity of meter readings must be protected. Lu et al. [144] proposed an efficient and privacy-preserving aggregation scheme. Each consumer's meter report and the aggregation report are signed by Boneh–Lynn–Shacham (BLS) short signature [147] which is secure under the computational Diffie–Hellman (CDH) problem in the random oracle model [148] Thus, integrity is guaranteed.

3.2.3   Authentication

Authentication implies that a receiver acknowledges that a received data packet is indeed from a claimed sender. A typical application is smart home network (SHN) [149–151], as depicted in Fig. 3.4. An SHN consists of a central control system, an outdoor user, several in-house electric appliances, and a cloud server. SHN provides many convenient and interesting services since it devices in SHN can communicate and this enables the user to remotely control the appliances when they are away

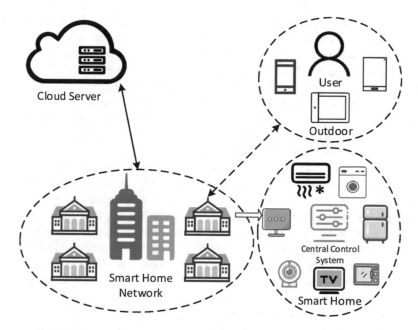

Fig. 3.4 Architecture of a typical smart home network

from home. For instance, the water heater can be configured before a user gets off work.

With the ubiquitous deployment of SHN, people are able to enjoy more modern home services. At the same time, it comes with the importance of authentication. Imagine the appliances can be controlled by any malicious entity without authenticating the instruction sender, all the appliances will turn into a chaos.

3.3   Privacy Concerns in Internet of Things

In this section, we introduce privacy concerns and corresponding countermeasures in different IoT scenarios. Privacy concerns mainly have identity, location, trajectory, report, and query.

3.3.1   Identity

Identity is the first priority of privacy protection. This is because people in the middle of some activities are mostly concerned about whether others will know who they are. These activities are sometimes sensitive, such as voting for a politician [152], purchasing a medical insurance [153], or being in a blind date [154].

The most common way to protect identity is for users to hide their identity and use pseudonyms [155–157] to participate in a system instead. Pseudonyms could be randomly chosen strings or a hash value of a real identity concatenated with a timestamp. However, the use of pseudonyms will make authentication a challenging task because the authenticator has no clue about who the real user is. Two methods solving this problem are pseudonym certification authority (PCA) [155] and anonymous authentication [158].

The first one creates anonymized temporary credentials as pseudonyms to users [155]. These credentials are used to cryptographically prove the qualification of a user when she/he is requesting to join in a system submitted samples and the qualification of users' submitted data. Since a reutilization of the same credential will help an adversary link the user, each user can obtain multiple credentials from the PCA. The second one provides anonymity for signers in a group while being able to revoke a signing key in case a user has misbehaved. Any signer in the group can sign messages but the identity of the signer is kept secret. However, the pseudonym changing strategy matters as well. For example, four vehicles are running on a street and if only one of them chooses a new pseudonym in the next time period, an adversary can easily record link between pseudonyms. Even though the four vehicles can change their pseudonyms at the same time, the speed and location embedded in transmitted messages can assist the adversary to link the pseudonyms, thus violating their privacy. Social spots are the places where several vehicles temporarily gather, e.g., the road intersection when the traffic light turns red or a free parking lot near a shopping mall. If all vehicles change their pseudonyms before leaving the spot, the first safety message that is broadcast includes indistinguishable information Location = social spot, Velocity = 0, and unlinkable Pseudonym. Then, the social spot acts as a natural mix zone to protect vehicles' identities.

Therefore, Lu et al. [159] propose that vehicles change their pseudonyms at social spots to break the link between identities and pseudonyms since a social spot is a natural mix zone where all vehicles change pseudonyms before leaving the social spot with indistinguishable information $speed = 0$ and $location = socialspot$.

3.3.2 Location

Location is the geographic position of a person. It has been extensively used in location-based services (LBSs), such as social rendezvous [160], traffic monitoring [161], location publication [162], map services [163]. These services need users' locations in order to guarantee normal operation, but location access has incurred new privacy threats to users. These threats range from localizing, profiling, and identifying a user. In addition, location data may reveal some sensitive information of users, such as home address, work address, health condition, and insurance status.

Location privacy protection mechanisms (LPPMs) are classified into four types.

Cloaking Cloaking offers location k-anonymity [164], i.e., a user cannot be distinguished from other $k - 1$ users. It hides a user's exact location into a wide spatial area to protect user location privacy. The size of the area could be adjusted according to user's specified privacy requirement. Zhu et al. [119] presented an anonymous smart-parking and payment scheme for vehicular networks to ease the public parking problem via utilizing private parking spots. Specifically, they divide the map of Beijing into a set of neighboring cells. When a cruising driver is looking for a parking spot, they expand the driver's location into a grid number to protect her/his location privacy. Then the driver sends this grid number instead of true location to a cloud server. They also mentioned that the cloaking size is an application and experience-based value. While cloaking technique can protect users' true locations, but it still exposes which area the user is located in so as to provide services. Therefore, there will always be a trade-off between location privacy and service utility.

Obfuscation Obfuscation alters users' locations and allows LBS servers to conduct some mathematical operations on the obfuscated locations. Ardagna et al. [302] presented several obfuscation operators to turn users' locations into circular areas such that LBS servers cannot identify a user's true location. The radius and center of the circle can be changed by the obfuscation operators. Duckham et al. [303] argue that obfuscation degraded the service quality within a pervasive computing environment in order to protect the privacy of the individual to whom that information refers. They designed a formal framework to provide obfuscated location-based services which achieved a balance between location privacy and high-quality services.

Dummy Dummy-based approaches require users to send many dummy locations to cover her/his true location [165] and location privacy is preserved since service providers cannot differentiate the true location. Shankar et al. [166] proposed a k-anonymization-based scheme to privately query LBSs where a user is sending a query as well as $k - 1$ sybil queries. These sybil queries include locations that are similar to the user's true location. Lee et al. [167] designed a dummy-based method for route services. When a user is about to query a route from a to b, she sends two sets A and B with $a \in A$ and $b \in B$ to a map server. The server calculates a route from each location in A to each location in B and returns all routes to the user. In order to reduce the computational costs, they designed an algorithm for the map server to calculate shortest routes between two groups of locations.

False Locations In false location-based methods, users only send fake locations other than a true location to a LBS server, and construct the right answers based on the results returned by the server. The advantage of such methods is that they can be easily integrated with existing LBSs. Yiu et al. [168] proposed a query framework in which points-of-interest (POIs) are first retrieved from the server incrementally. The process begins with a location different from a user's true location and retrieved the nearest neighbors. It ends until an accurate query result can be found. Peddinti et al. [169] presented a location privacy protection scheme to let the user have LBS without exposing her/his true location. Users do not use the true location to send a query to a LBS server, but queries a result from some nearby locations, i.e., cover locations. The LBS server returns corresponding results to the user who processes locally to obtain a result related to the true location.

However, users have to continuously query the LBS servers and then obtain enough related results to acquire accurate results which leads to a huge response delay.

3.3.3 Trajectory

Trajectory, like users' mobility traces, has been extensively applied in many applications, such as route choosing [170], urban planning [171], and market analysis [172]. Trajectory publication services make it easy for the public to acquire real-time trajectory information and analyze travel patterns. However, users consider where they travel from and where they are going as a part of their privacy [173], since these trajectory data can be used to infer detailed activities and predict next movements.

Chen et al. [174] firstly introduced a variable-length n-gram model to achieve differential privacy [175, 176] for publishing sequential data and developed several techniques (i.e., how to allocate privacy budget, how to choose a threshold value, and how to enforce consistency constraints) so as to guarantee system utility. Li et al. [177] presented a new novel trajectory data publication algorithm based on differential privacy [178]. Specifically, they first merge close trajectories of users and construct a new trajectory dataset. Then they iteratively generate bounded Laplace noises. The noise generation algorithm is elaborately designed with regard to the count constraint of each road.

3.3.4 Report and Query

Li et al. [177] aim to protect query and report privacy in participatory sensing. Nowadays, the ubiquity of smart devices has paved the way for achieving a large-scale sensing and enabled people to collect and share information in their surroundings as depicted in Fig. 3.5. But users will inevitably contain sensitive information in their data reports and queries if they participate in the sharing system. For example, if a user continues to query a LBS server about popular restaurants and entertainment venues within a certain region, then it is of great possibility to infer that the user lives or works in this area. Therefore, an appealing query and report privacy protection mechanism is needed to increase the users' participation and ensure the durability of the participatory sensing system.

Li et al. [179] proposed a query and report privacy-preserving scheme for users in crowdsensing systems [180, 181]. Specifically, users (i.e., reporters and queriers)

Fig. 3.5 Architecture of
crowdsensing

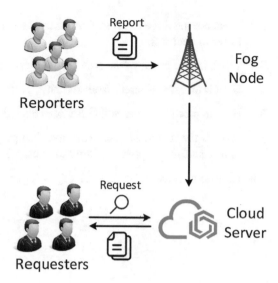

first register to a trusted authority with their interested tags (e.g., gas station, ATM) and then the trusted authority returns a pair of keys to queries and a public key with a tag key to reporters. Then users encrypt reports or queries with a tag key, and send the ciphertexts to a server which will match different reports and queries. Finally, reports will be returned to corresponding queriers by the server.

3.4 Summary

This chapter discusses security issues and privacy concerns as well as their corresponding countermeasures in IoT systems. Although existing work has been devoted a lot into mining security problems and designing protection mechanisms, we are facing many unexploited challenges in high-growth IoT systems. More efforts are required to locate these challenges in time and treat them as opportunities of building a more secure and reliable environment.

Exercises

3.1 Please explain confidentiality, integrity, and authentication. Refer to Sect. 3.2.

3.2 Say Alice and Bob are transmitting messages using a symmetric encryption scheme, what are the four attacks that an adversary Eve can launch to break the scheme?

3.3 What are the three attacks of an adversary trying to break a signature scheme? Refer to Sect. 3.2.

3.4 What are the privacy concerns in IoT application? Refer to Sect. 3.3.

3.5 How is privacy different from security? Refer to Sect. 3.3.

3.6 How to protect user identity? Refer to Sect. 3.3.

3.7 What is the relationship between location privacy and trajectory privacy? How to protect the two privacy? Refer to Sect. 3.3.

3.8 How do adversaries violate privacy in report and privacy? Refer to Sect. 3.3.

Part II
Blockchain in Privacy-Preserving Cloud Data Storage Services

Chapter 4
Blockchain-Enabled Cloud Data Preservation Services

4.1 Overview

Cloud data preservation [182] is becoming an indispensable part of our daily life. The traditional cloud records which are in the form of paper have several disadvantages [183], such as easily lost or damaged. There is a strong need for a quick transition for electronic data. For example, modern hospitals are encouraged to utilize electronic records instead of paper records [184]. Meanwhile, cloud data are prone to network attacks [185] and it is necessary to protect these data by data preservation technologies.

Some hospitals record the whole process of treatment by electronic data and store cloud records in their databases. By doing so, a database manager could extract the required data from the database when a doctor queries a prescription record [186]. However, a patient's real health condition cannot be accurately provided if the cloud data are be falsified or deleted. When the data are preserved by a third party, the privacy of patients would be leaked, and it is even more challenging to difficult to guarantee reliability and availability. Blockchain is a public distributed ledger built on P2P networks and group consensus algorithms [187]. It has a growing list of linked blocks, which ensures that every data record in blocks will be stored permanently and cannot be tampered with. Although blockchain is a good choice to solve the cloud data preservation problems, still there are some challenges. For instance, the volume of cloud data takes too much storage space [188] and how to prove that preserved cloud data have not been tampered with.

This chapter proposes a new blockchain-enabled cloud data preservation (CDP) system which provides a reliable cloud data preservation solution to protect the data from falsification, efficiently verify the data validity, and preserve user privacy. Motivated by the *proof of possession of balance* [189], a concept of *proof of primitiveness of data* is proposed. It means that the data will be stored on the blockchain for good. There is no need for us to worry about data falsification since the system can verify whether the retrieved data are the same as the original data.

© Springer Nature Switzerland AG 2019
L. Zhu et al., *Blockchain Technology in Internet of Things*,
https://doi.org/10.1007/978-3-030-21766-2_4

Specifically, this chapter presents a solution of the CDP system [12], which uses a hybrid cryptosystem to encrypt users' cloud data to protect their privacy. Then, it adopts different storage schemes for cloud data with different data formats. For example, textual data can be directly stored in blockchain, and files with a large amount of data are distributed stored in the form of sharding. Next, the concept of *proof of primitiveness of data* is presented, which ensures that the data is not tampered with and will never lost. Finally, experimental results show that the storage costs are acceptable.

The remainder of this chapter is organized as follows: Sect. 4.2 presents the basic techniques of cloud data preservation services. Then, the technical dimensions in cloud data preservation services are introduced in Sect. 4.3. Section 4.5 presents a solution of blockchain-assisted cloud data preservation services in Sect. 4.4 and provides the use case in Sect. 4.5. Lastly, a summary is drawn in Sect. 4.6.

4.2   Technical Dimensions in Cloud Data Preservation Services

4.2.1   Essential Components

The essential components in cloud data preservation services include clients, data access applications, and blockchain network which are depicted in Fig. 4.1.

A typical cloud data preservation operation flow includes the data access program and the blockchain interaction program. The data access program has four operations: data submission, data manipulation, data query, and data verification. Clients upload cloud data for preservation and query the preserved cloud data. In the blockchain interaction program, the system stores cloud data on the blockchain and extracts the stored data back to the data access application. Cloud data is transmitted to the blockchain network in the form of transactions. The greater the amount of cloud data, the larger the reward for miners to receive.

4.2.2   Threat Model

Data Modification or Deletion   We consider that the CDP system preserves cloud data including diagnosis certificates and cloud records. It is possible for criminals to deliberately modify or even delete cloud data [190, 191].

Fraud by Using False Data   Cloud data could be served as a proof of rights or judicial credentials, but how to identify the real one is crucial when there are two pieces of different or conflicting cloud data.

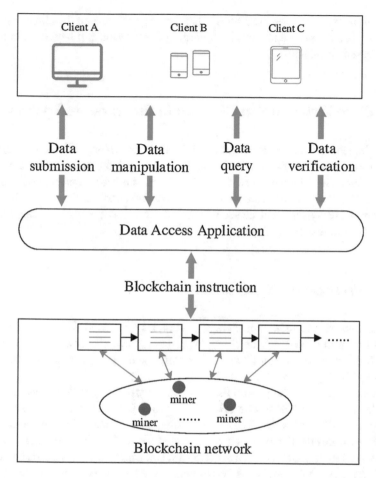

Fig. 4.1 Architecture of cloud data preservation

Privacy Violation Adversaries can download all cloud data and make use of them in a malicious way. For example, adversaries can know one patient's physical conditions from the cloud records, thereby violating their privacy [192].

4.2.3   Design Goals

- **Data consistency**. Clients' cloud data need to be stored on the CDP server but some data packets may be lost during data transmissions, and subject to malicious tampering [193]. In order to ensure that the data submitted by the user are the same as the data the CDP server receives, the system should guarantee data consistency.

- **Anonymity**. The preserved cloud data are highly related to clients' privacy. Cloud data must be anonymized, including the relationship between clients and preserved cloud data.

4.3   Basic Techniques in Cloud Data Preservation Services

We implement two contracts, which are PS (*Preservation Submission*) and PV (*Primitiveness Verification*). In the PS phase, clients upload their data to the CDP system. The data will be stored in the CDP system after correctness checking. In the PV phase, clients can view the stored data and continue to upload new data. The CDP system will check whether the requested data has been changed and returns the results to the user.

4.3.1   Preservation Submission

PS has three steps: $dataProcessing()$: receiving the cloud data uploaded by the clients user and processing the data; $dataProtection()$: encrypting data, and the $writeInBlockchain()$: storing the preservation on the blockchain. We describe the three steps in Algorithms 1–3.

Algorithm 1 states the pseudo code of $dataProcessing()$ that processes user's uploaded data. The CDP system can handle two data formats: file and text. For the first type, CDP chooses the SHA256 algorithm to calculate a hash value $hash_r$ of the file and stores the file in randomly generated file folders. For the second type, the text is stored in the database for a short period and a similar hash value is calculated. Lastly, the ECC algorithm is invoked to generate a pair of asymmetric keys.

Algorithm 2 states the pseudo code of $dataProtection()$ that encrypts the above processed data. The CDP system chooses the AES algorithm to create a symmetric key K_{sym} in order to encrypt the unpreserved data. For a file, the content M to be encrypted is a location index of the file $index$. Content M is encrypted to be C:

Algorithm 1 $dataProcessing()$

Require: The cloud data, client information
 1: **if** Format of data is file **then**
 2:      Generate random folders
 3:      Store the file
 4:      $hash_r \leftarrow$ calculate a hash value of the file
 5: **else if** Format of data is text **then**
 6:      Store the text in database temporarily
 7:      $hash_r \leftarrow$ calculate a hash value of the text
 8: **end if**
 9: Generate a pair of keys

Algorithm 2 $dataProtection()$

Require: Client's query
Ensure: Encrypted preservation
 1: **if** The client confirms the preservation **then**
 2:      Generate a symmetric key K_{sym}
 3:      $C \leftarrow$ Use K_{sym} to encrypt the data to be preserved M
 4:      $pubKey \leftarrow$ Get the public key generated in Algorithm 1
 5:      $En_r \leftarrow$ Use $pubKey$ to encrypt K_{sym}
 6:      **return** C
 7: **else**
 8:      Delete the temporary storage of data preservation
 9:      **return** Null
10: **end if**

Algorithm 3 $writeOnBlockchain()$

Require: The encrypted preservation
Ensure: Hash value of blockchain transaction
 1: **if** The identity of the user is valid **then**
 2:      Store data E in the blockchain
 3:      $Tx \leftarrow$ Get hash value of blockchain transaction
 4:      **return** Tx
 5: **else**
 6:      **return** Null
 7: **end if**

$$C = AESEnc(K_{sym}, M). \tag{4.1}$$

Then, $pubKey$ generated in Algorithm 1 is used to encrypt K_{sym}. Finally, the algorithm sends back the encrypted data.

Algorithm 3 states the pseudo code of $writeInBlockchain()$ that writes the preservation on the blockchain. If the user is a legal user, his/her data will be written on the blockchain and a transaction hash value Tx will be sent back to the user. Otherwise, the CDP system declines to process the data.

4.3.2   Primitiveness Verification

PV consists of three steps: (1) $checkPreservedData()$: viewing the preserved data; (2) $validatePreservedData()$: verifying the consistency with the original data; (3) $getChainData()$: extracting data from the blockchain. We describe the three steps in Algorithms 4–6.

Algorithm 4 states the pseudo code of $checkPreservedData()$ which reviews the preserved data in the blockchain. The CDP system retrieves the preserved data C from the blockchain according to Tx. Then, $priKey$ generated in Algorithm 1 is used to decrypt C.

Algorithm 4 $checkPreservedData()$

Require: Private key, hash value of blockchain transaction Tx
Ensure: Decrypted data
 1: $E \leftarrow$ Extract data from blockchain based on Tx
 2: $C \leftarrow$ Extract the data from E
 3: $priKey \leftarrow$ Get the private key generated in Algorithm 1
 4: $M \leftarrow$ Use $priKey$ to decrypt En_K to decrypt C
 5: $hash_r \leftarrow$ Extract the hash from E
 6: $hash_c \leftarrow$ Calculate the hash of M
 7: **if** $hash_r$ is the same as $hash_c$ **then**
 8:     **return** M
 9: **else**
10:     **return** Null
11: **end if**

Algorithm 5 $validatePreservedData()$

Require: Data to be verified, blockchain transaction hash Tx
Ensure: A Boolean value
 1: $hash_c \leftarrow$ Calculate the hash of the data U
 2: $E \leftarrow$ Extract data from blockchain based on Tx
 3: $C \leftarrow$ Extract the encrypt data from E
 4: $hash_r \leftarrow$ Extract the hash from E
 5: **if** $hash_r$ is the same as $hash_c$ **then**
 6:     **return** TRUE
 7: **else**
 8:     **return** FALSE
 9: **end if**

$$M = AESDec(ECC(priKey, En_K), C). \tag{4.2}$$

Next, the stored $hash_r$ is retrieved and compared with a newly calculated hash value $hash_c = SHA_{256}(M)$ of the decrypted preserved file M. If $hash_r$ is the same as $hash_c$, then M has not been modified, and the decrypted preservation is sent back to the user. Otherwise, null is returned.

Algorithm 5 states the pseudo code of $validatePreservedData()$ that receives the data uploaded by a user and checks whether the data are the same as the preservation. The user uploads the data U to the CDP system which calculates a $hash_c = SHA_{256}(U)$. Then, the data stored in the blockchain are retrieved including a hash value of the original data $hash_r$. If $hash_c$ is the same as $hash_r$, then U is the same as the data preserved, and true is returned. Otherwise, false is returned.

Algorithm 6 states the pseudo code of $getChainData()$ that extracts and processes data stored in the blockchain. If Tx exists and it is a valid, the DPS system retrieves the whole transaction and extracts its data. The data are recoded and parsed into the original format, and returned. Otherwise, null is returned.

Algorithm 6 $getChainData()$

Require: The hash of blockchain transaction Tx
Ensure: The parsed data
1: **if** Tx is valid **then**
2:      Obtain information contained in the transaction based on the hash value
3:      Extract the stored data
4:      Recode and parse
5:      **return** Parsed data
6: **else**
7:      **return** Null
8: **end if**

4.4 Solution

This section presents a solution of blockchain-assisted cloud data preservation system CDP which mainly has two phases. The first phase commits unpreserved data when the user wants to secure the new data. The second phase identifies the primitiveness of the data. The information flow of the CDP is shown in Fig. 4.2.

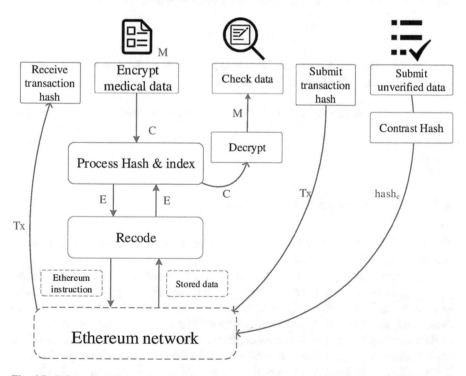

Fig. 4.2 Information flow of the CDP system

4.4.1 Data Submission

The data submission phase includes data submission and processing, data validation, and data preservation.

The legal users detected can submit cloud data to the CDP system. Specifically, the users invoke the $dataProcessing()$ procedure to save the data in a newly created folder or a database according to the data type. Then they generate a hash value $hash_r$ of the data. To prevent data loss, the CDP system splits data into files of size 1MB and randomly distributes the files into different folders.

Based on the submitted data, a feedback of the data is generated and a hash value is computed as a confirmation. Next, the $dataProtection()$ procedure is invoked. The public-key encryption algorithm we use is elliptic curves cryptography. The generated private key $priKey$ will be sent to the client's designated mailbox such that the client can decrypt and view the private data.

The $writeInBlockchain()$ procedure is invoked to write the data in the blockchain. Because the size of the available file that can be written in the $Ethereum$ blockchain is constrained, the contents cannot be written fully. If the data that need to be preserved are text files, the $hash_r$ and encrypted data C_t will be combined to E and then written in the blockchain directly.

$$E =< hash_r, C_t > . \tag{4.3}$$

If the data are multimedia files, the encrypted index of the data location and the hash value $hash_r$ will be stored in the blockchain.

$$M_f =< index_i > (i = 1 \rightarrow n, n = files'\ amount). \tag{4.4}$$

$$E =< hash_r, C_f > . \tag{4.5}$$

4.4.2 Primitiveness Identification

The primitiveness identification phase includes data acquisition of preservation contents and verifying consistency.

The client can directly view the contents of previously preserved data by providing the private key $priKey$. First, the $getChainData()$ procedure is invoked to extract the data from the blockchain and divide the $hash_s$ and the encrypted data. Next, the client uses $priKey$ to decrypt the data to obtain the stored data or data location index. The text data is directly obtained from the blockchain. The multimedia data require the reassembly of the stored files based on the location index, followed by the computation of the $hash_c$. If $hash_s$ is the same as $hash_c$, the client will acknowledge that the data contents have not been falsified.

Clients can verify whether the local data are exactly the same as the preserved data. The client uploads the data which needs validation. Then, the CDP system invokes the *validatePreservedData*() procedure to compute the $hash_c$, extract the data from the blockchain, and divide the stored $hash_r$, which is compared with the $hash_c$ of the data to be validated. If the procedure outputs *True*, the data consistency is guaranteed. Otherwise, the CDP system will notify clients of the inconsistency.

4.5   Use Case

In this section, we evaluate the CDP system in terms of its operational costs. The cost is measured by the total *Gas* and it will be incurred only when the data are stored. The amount of *Gas* will be determined when an *Ethereum* transaction is produced, e.g., a balance operation costs 20 *Gas* [194]. We conduct experiments on a MacBook laptop with macOS 10.11, 2.2 GHz quad-core Intel Core i7 CPU, 16 GB RAM. We implement the CDP system based on the *Ethereum* and *geth* [195]. The total *Gas* amount is transformed into USD by referring to the exchange rate chart *coinbase* [196] and *myetherwallet* [197].

A basic transaction requires 21,000 *Gas*. For every 100 bytes text or 10 MB file added, it costs about 5400 *Gas*. Figure 4.3 shows the total operational cost. Because the *Gas* cost is determined by the length of the data, the cost increases with the size

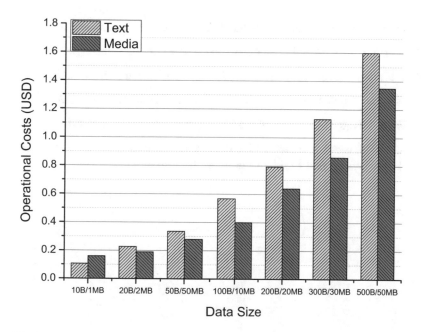

Fig. 4.3 Operation cost for data preservation under different data types

of the data. However, the total operational cost is still very low and even if the size of the multimedia file is 50 MB or the number of words is 500 bytes, the total cost is less than US$2.

4.6   Summary

The CDP scheme provides a reliable cloud data preservation solution to ensure the primitiveness and verifiability of cloud data and preserve user's privacy. Compared with existing data preservation systems, a promising feature of CDP is that it can deal with scenarios where cloud data are lost and falsified. Even if the data are not the same as the original form, they can be retrieved and verified through the blockchain. We also implemented the CDP system based on the *Ethereum* platform. The experimental results show that the costs of preservation with the CDP system are less than US$2, even if the size of the preserved files is 50 MB. When the number of concurrent preservations is 400 and all files are 30 MB, the response time of the system is not less than 1 s.

Exercises

4.1 Please give some examples for the data type in CDP services. Refer to Sect. 4.2.

4.2 What are the essential components in CDP services? Refer to Sect. 4.2.1.

4.3 What is the security model in the CDP system? Refer to Sect. 4.2.2.

4.4 What are the design goals in the CDP system? Refer to Sect. 4.2.3.

4.5 What are the main phases of the CDP system? Refer to Sect. 4.4.

4.6 Say Alice wants to transfer a file to Bob with an evidence that acknowledges this behavior, what could they do with the help of a blockchain?

Chapter 5
Blockchain-Enabled Controllable Data Management

5.1 Overview

The recent development in blockchain gives the credit to the huge popularity of Bitcoin. The blockchain's capability to offer a transparent data utilization and sharing opportunity [198, 199] has cultivated an increasing number of blockchain-based schemes [200–205] have been proposed. A blockchain is deemed as a reliable platform and all transactions are recorded in blocks that are chained together via cryptographic hashes. The growing chain makes it computationally difficult to alter an existing block without being detected. Blockchain allows flexible access and it could be adopted to realize privacy-preserving property, e.g., through the use of anonymous accounts [8, 118, 206, 207]. Some cryptocurrency systems protect the real identity of users by permitting them to register an anonymous account [208].

This chapter presents a new method [209] to address the control issue in blockchain by proposing a blockchain-based controllable data management (BCDM) model (Fig. 5.1). Specifically, a trusted authority (TA) node in the network system is introduced and the voting authorization level of TA is higher than other nodes. TA is configured with a veto power, which is delicately designed for resisting malicious voting. In addition, the BCDM model utilizes a cloud server to improve storage efficiency (i.e., storage-as-a-service (StaaS)). Each block stores only metadata so as to optimize the efficiency of block construction and reduce storage waste.

The remainder of this chapter is organized as follows: in Sect. 5.2, we present the technical dimensions in controllable data management services. We introduce the basic techniques of controllable data management services in Sect. 5.3. We present an advanced solution in Sect. 5.4 and provide the use case in Sect. 5.5. Lastly, we draw our summary in Sect. 5.6.

© Springer Nature Switzerland AG 2019
L. Zhu et al., *Blockchain Technology in Internet of Things*,
https://doi.org/10.1007/978-3-030-21766-2_5

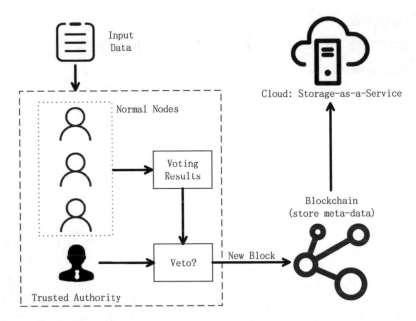

Fig. 5.1 High-level architecture

5.2   Technical Dimensions in Blockchain-Based Controllable Data Management

5.2.1   Essential Components

The BCDM system model consists of four entities: trusted authority (TA), cloud servers (CS), blockchain system (BS), and a set of users $U = \{U_1, U_2, \cdots, U_N\}$ which are depicted in Fig. 5.2. We consider the possibility that some users can request TA to revise documents $D = \{D_1, D_2, \cdots, D_M\}$ and send other users the revised documents $nD = \{nD_1, nD_2, \cdots, nD_S\}$, where each $nD_i \in nD$ is a new version document. Other users participate in voting for the revised documents in the document management system.

- **Trust Authority (TA)**: TA is an authoritative entity in blockchain system and it could examine the revised document to determine whether it is valid and drop it on its own.
- **Cloud Servers (CS)**: Cloud servers store the encrypted revised documents. It is helpful for TA and users to check the revised documents without making the same revision on documents.
- **Blockchain System (BS)**: Blockchain system collects the votes from users for revised documents to determine whether users acknowledge the revision. There are three sub-entities, which are a system administrator SA, a set of voting contracts, and a set of counting contracts. SA registers users and handles

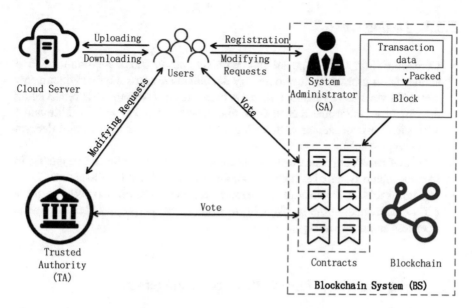

Fig. 5.2 The system model of blockchain-based controllable trustworthy document management scheme

their requests to revise documents. SA distributes keys and address to users and key pairs to voting contracts $VC = \{VC_1, VC_2, \cdots, VC_R\}$ and counting contracts $CC = \{CC_1, CC_2, \cdots, CC_R\}$. Voting contracts verify users' identities and provide them a platform to vote. Counting contracts count the number of votes and make the votes and results public. Blockchain stores records of revised documents $cR = \{cR_1, cR_2, \cdots, cR_T\}$ and voting information $vR = \{vR_1, vR_2, \cdots, vR_Q\}$ and the hash of revised documents $dH = \{dH_1, dH_2, \cdots, dH_T\}$. Blockchain puts the data mentioned above into blocks when these records are published. Users can review the blocks to retrieve the hash of previously revised documents and check the previous revisions to documents.

- **Users** ($U = \{U_1, U_2, \cdots, U_N\}$): A $U_i \in U$ registers to SA and revises documents by sending requests to SA and voting contracts. Each user then has the ability to decide whether to acknowledge the revisions of documents.

5.2.2 Threat Model

TA is a fully trusted entity. Users can collude with each other to revise documents in order to make documents invalid, then they vote to approve the revision such that the invalid documents are acknowledged. Some users may behave maliciously and they are recorded on the blockchain and reviewed by other users. Besides, external adversaries may try to forge some legal users' signatures so as to vote for illegal documents.

5.2.3   Design Goals

- **Controllability**: TA is a special node with veto power and it can monitor voting actions $\mathscr{A}$. A user who wants to modify documents D sends modified documents nD to TA via a secure channel, then TA compares nD with the data retrieved on the cloud by hash values dH to determine whether they are valid. Additionally, adversaries cannot tamper with voting record vR because they cannot decrypt the encrypted vR.
- **Privacy-Preserving**: Each user needs to obtain the permission $\mathscr{P}$ of the TA in the blockchain system. No user can obtain other users' real identities or vR.
- **Openness and Transparency**: Both modification records cR and voting records vR are published on the blockchain for users' convenience in order to provide openness and transparency.

5.3   Bridging Blockchain with Data Management

Revising documents is a common operation in document management, but it is crucial to guarantee the legality and security of revised documents in a data management system (DMS).

For instance, revising architectural effect blueprints is familiar in the construction industry. Revising a document without any validation will incur inconsistency among all architects. Therefore, a transparent, trustworthy, and traceable method is important for validating document revisions. Here, a blockchain-based system can be used to record users' revision requests. A special node TA is also introduced to monitor user behaviors.

5.4   Solution

There are three phases in the BCDM scheme: system initialization, document modification, and document management.

5.4.1   System Initialization

This phase initializes system parameters and generates key for user registrations.

1. **Setup:** SA chooses a security parameter λ to generate system parameters $\{G_1, G_2, \hat{e}, q, P, H\}$, where G_1 is a multiplicative group, G_2 is an additive group, q is a λ-bit prime number and the order of G_1 and G_2, $\hat{e}$ is a bilinear pairing, P is a generator, and H is a hash function. SA generates a pair of keys

G_{pk}, G_{sk}, and generates keys pairs to contracts. TA generates a pair of keys. All addresses are generated from public keys.

2. **Registration:** SA generates public keys $U_{pk} = \{U_{pk1}, U_{pk2}, \cdots, U_{pkN}\}$ and private keys $U_{sk} = \{U_{sk1}, U_{sk2}, \cdots, U_{skN}\}$, signature keys $U_{sig} = \{U_{sig1}, U_{sig2}, \cdots, U_{sigN}\}$ for registering users.

5.4.2 Document Modification

This phase includes several operations of the voting implementation.

1. **Request**: A user encrypts a request message $\mathcal{M}$ to revise documents under private keys and sends it to SA. U_i also sends a signature key and the revised document encrypted by a private key to TA through a secure channel.
2. **Verification**: SA decrypts the encrypted request to obtain $\mathcal{M}$ and verifies the signature key. If it is valid, SA acknowledges the user's revision request and makes it a voting option.
3. **Vote**: The user computes a hash value of voting option and sends it to TA. TA generates several signature parameters and distributes them to users. Then user computes a signature for voting option and sends the signature to a voting contract. The voting contract verifies the signature. If the signature is valid, voting contract signs it and returns it back to user. User encrypts the signed signature, signature, and voting option under a counting contract's public key and sends them to the counting contract. TA with the veto power right also votes for revised documents.
4. **Count**: If the signatures are valid, counting contract will record the vote and finally computes the voting results. If TA votes for a decision, the counting contract publishes the voting result based on TA's vote.

5.4.3 Document Management

This phase includes several primary operations on blockchain.

1. **RecordStore**: Records about revised documents, votes, and TA's decisions are packed in blocks with hash values of revised documents. Users and TA can verify the revised documents and find the documents exactly through the corresponding hash values.
2. **Upload**: Users upload revised documents and all members can decrypt the documents under secret keys.
3. **Download**: Users and TA have an option to download revised documents. TA can compare the downloaded documents with the ones from users and check whether they are the same which is a reference for request rejection.

5.4.4   User Registration

If users participate in the blockchain system, they have to obtain identities and signature keys in order to revise documents and vote for document revisions.

SA generates keys set U_{pk}, U_{sk} for users **U**. Blockchain system generates users' addresses set U_{Addr} based on U_{pk}. SA and TA randomly choose number $(x_i{}^u, y_i{}^u)$ to create signature keys set U_{sig} for each user. SA computes $x_i{}^s = (x - x_i{}^u) \bmod q$ and sends $(x_i{}^s, U_i)$ to TA. TA computes $y_i{}^s = (y - y_i{}^u) \bmod q$ and stores $y_i{}^s$. When users vote for a revision request, TA checks whether their identities are valid by computing $x_i{}^s$, $x_i{}^u$, $y_i{}^s$, and $y_i{}^u$.

5.4.5   Voting and Counting

For each voter, she sends her vote $H(V_i)$ and a signature key U_{sig_i} to TA, TA verifies this voter. Some detailed requirements are:

- Each user's signature should be different from others' signatures: $x_i{}^u \neq x_j{}^u$ and $y_i{}^u \neq y_j{}^u$, where $i \neq j$.
- Any random numbers cannot add up to be the same as x and y: $x_{i_1}{}^u + x_{i_2}{}^u + \cdots + x_{i_j}{}^u \neq x \bmod q$ and $y_{i_1}{}^u + y_{i_2}{}^u + \cdots + y_{i_j}{}^u \neq y \bmod q$, where $j \in \mathbb{Z}$.
- Any random numbers cannot add up to be the same as $x_j{}^u$ and $y_j{}^u$ to avoid $x_j{}^u$ and $x_j{}^u$ being guessed: $x_{i_1}{}^u + x_{i_2}{}^u + \cdots + x_{i_j}{}^u \neq x_j{}^u \bmod q$ and $y_{i_1}{}^u + y_{i_2}{}^u + \cdots + y_{i_j}{}^u \neq y_j{}^u \bmod q$, where $j \in \mathbb{Z}$.

After TA verifies the signature keys, it computes a part of signatures $(v_i{}^s, \sigma_i{}^s)$, where $v_i{}^s = y_i{}^s * H(V_i)$, $\sigma_i{}^s = x_i{}^s * H(V_i)$. Then, TA stores $(U_{Addr}$ and $H(V))$. Voters compute full signatures by the random number $x_i{}^u$.

Each voter sends to voting contract a signature $\sigma_i = v_i{}^s + v_i{}^u + \sigma_i{}^s + \sigma_i{}^u$. Voting contract verifies the signatures. $v_i{}^u = y_i{}^u * H(V_i)$ and $\sigma_i{}^u = x_i{}^u * H(V_i)$. After voters' signatures are verified, voting contract signs voters' signatures under private key Y_A.

Voters encrypt $\{V_i || Y_A(\sigma_i) || \sigma_i\}$ under public key X_B and sends $Enc\{V_i || Y_A(\sigma_i) || \sigma_i\}$ to counting contract. Counting contract decrypts $Y_A(\sigma_i)$ under public key X_A to verify the signature. Counting contract calculates a hash value of V_i to validate the vote. After verifying voters, counting contract records the voting information according to $H(V_i)$.

If TA votes for a decision, counting contract sets the voting result $\mathscr{R}$ to be TA's vote. If TA approves a decision, blockchain system calculates the document hash value. Otherwise, "Denied" will be the document hash value.

5.5 Use Case

5.5.1 Experiment Evaluation

This section analyzed the cost and efficiency of the BCDM system. Cost refers to the financial cost in the mining process and efficiency is the time of constructing blocks. We used an Ethereum client Geth and an Ethereum Wallet running on a computer (macOS 10.13.4, 2.3 GHz Intel Core i5, 8 GB of 2133 MHz LPDDR3). 0.018 Ether is configured for a million gas. Table 5.1 records experiment settings regarding size of document and number of users.

The time includes time $t_{register}$ for user registration and time t_{vote} for user votes, and publishing transactions and hash values of modified documents. Figures 5.3, 5.4, 5.5, 5.6, 5.7, 5.8, 5.9, and 5.10 show the time costs for users in different documents. We can observe that the cost has a positive relationship with both number of users and size of document. The hash value of the document can be chosen by users through smart contract, and the advantage of this approach is that the size of the document has limited impacts on the efficiency.

From Fig. 5.11, we can see that the gas cost increases with the number of users. This is reasonable for the increase in the computational workload. Figure 5.12 shows that the number of the revised document is linear to the gas cost and the size of document has a strong impact on the time cost. Figure 5.13 illustrates that the time of packing documents by hash values is also linear to the number of revised documents.

5.6 Summary

This chapter presented a new method to achieve a blockchain-based controllable data management (BCDM) system. It addresses the control issue on the public ledgers. The introduction of a special node TA in BCDM enables one to abort any potentially malignant behaviors.

Table 5.1 Experimental settings

Setting	Size of document (MB)	Number of users
1	10	[0, 50]
2	20	[0, 50]
3	50	[0, 50]
4	100	[0, 50]
1	10	[0, 100]
2	20	[0, 100]
3	50	[0, 100]
4	100	[0, 100]

Fig. 5.3 Comparisons of time costs for users under setting 1

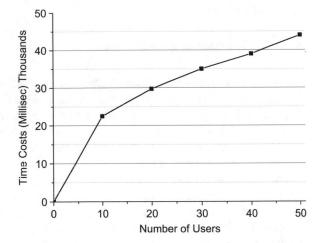

Fig. 5.4 Comparisons of time costs for users under setting 2

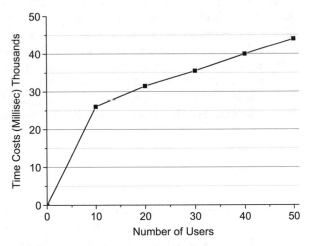

Fig. 5.5 Comparisons of time costs for users under setting 3

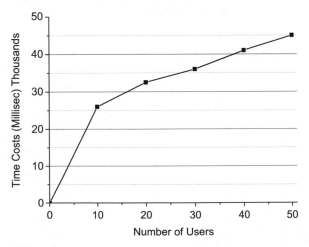

Fig. 5.6 Comparisons of time costs for users under setting 4

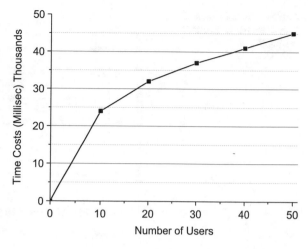

Fig. 5.7 Comparisons of time costs for users under setting 5

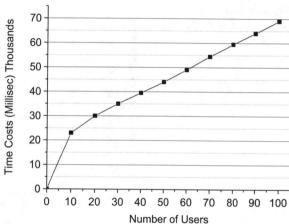

Fig. 5.8 Comparisons of time costs for users under setting 6

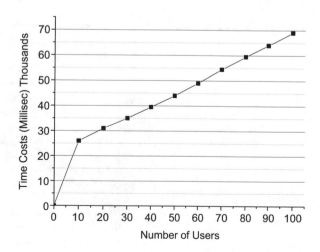

Fig. 5.9 Comparisons of
time costs for users under
setting 7

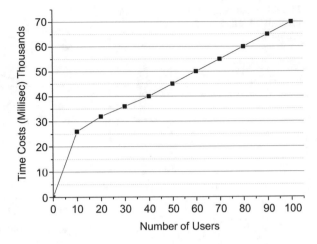

Fig. 5.10 Comparisons of
time costs for users under
setting 8

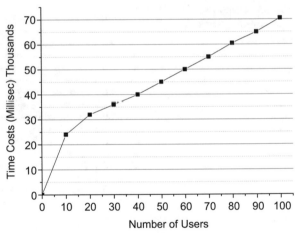

Fig. 5.11 Gas costs of votes

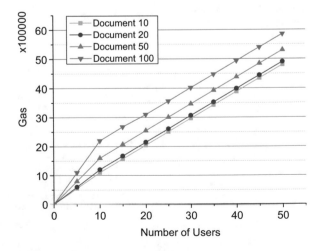

Fig. 5.12 Gas costs of packing hash values of documents

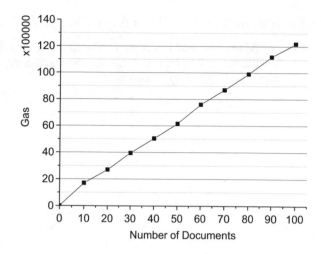

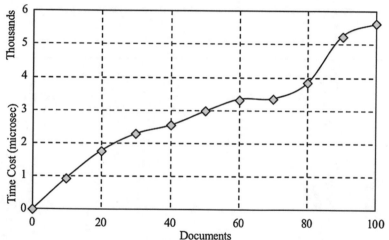

Fig. 5.13 Time cost for packing hash values of documents

Exercises

5.1 Please name a few scenarios where data needs to be controlled. Refer to Sect. 5.1.

5.2 What are the essential components in BCDM services? Refer to Sect. 5.2.1.

5.3 What is the security model in the BCDM system? Refer to Sect. 5.2.2.

5.4 What are the design goals in the BCDM system? Refer to Sect. 5.2.3.

5.5 What are the main phases of the BCDM system? Refer to Sect. 5.4.

5.6 Say Alice, Bob, and Cathy are conducting one ongoing project from different locations, and they have to guarantee the consistency of the project, what could they do with the help of a blockchain?

Part III
Privacy-Preserving Blockchain Technology in Internet of Things

Chapter 6
Blockchain-Enabled Vehicle Electricity Transaction Services

6.1 Overview

Vehicle-to-grid (V2G) networks, including electric vehicles (EVs), are gradually developing as an essential enhancement for smart grids. In V2G networks, two-way electricity transmissions create a huge number of electricity usage payment records. These records can be used to provide valuable services, such as electricity usage forecasting, electricity price estimation, and optimal electricity scheduling [210–212]. To put the benefits aside, the sharing of payment records will incur privacy concerns because sensitive information may be leaked. Such information include real identities, current locations, and charging/discharging volumes. Therefore, it is important to balance privacy and utility in V2G networks.

Traditional payment options (e.g., debit/credit cards and mobile payments) require interactions with third parties. These interactions bring about distributed payment records containing user information, which can be shared without consents from users. Anonymous payment schemes [213–215] enable privacy protections for EVs. Many of them adopt a centralized design and use a trusted third party to handle all payments. Although these schemes meet their proposed privacy requirements, they cannot protect identities when the payment records are shared among multiple entities for further analysis. Meanwhile, the centralized design is subject to the risk of a single point of failure.

As an uprising technique in digital currency systems [216], blockchain has advantages in record authenticity and distribution, which can be utilized in building an anonymous payment mechanism while enabling data sharing in V2G networks. There are two challenges to be tackled. The first is to simultaneously guarantee the reliability and efficiency of transactions. For example, Bitcoin only supports seven transactions per second and Hyperledger [217] records agreed-upon transactions

© Springer Nature Switzerland AG 2019
L. Zhu et al., *Blockchain Technology in Internet of Things*,
https://doi.org/10.1007/978-3-030-21766-2_6

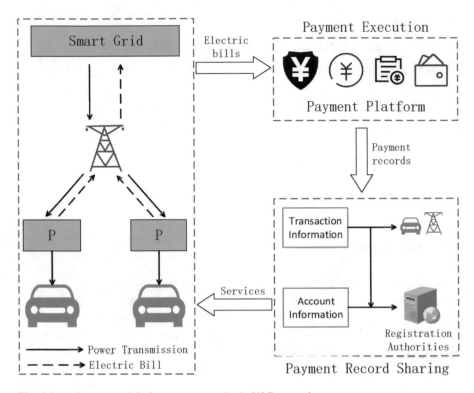

Fig. 6.1 Reference model of payment scenarios in V2G networks

without verifying transaction that sacrifices the reliability. The second is the auditability. Users are encouraged to generate pseudonyms so as to protect their real identities. This approach creates a barrier to identify malicious transactions (Fig. 6.1).

This chapter presents a blockchain-enabled vehicle electricity transaction scheme VET for V2G networks. Specifically, we establish a registration process to protect the traders' privacy while enabling payment auditing by privileged users. Then, we create a novel type of transaction structure and a corresponding transaction verification algorithm. Finally, we present a proof-of-concept prototype and analyze the feasibility and effectiveness.

The remainder of this chapter is organized as follows: in Sect. 6.2, we present the technical dimensions in vehicle electricity transaction services. We introduce a solution of blockchain-assisted vehicle electricity transaction services in Sect. 6.3 and provide the implementation scenario in Sect. 6.4. Lastly, we draw our summary in Sect. 6.5.

6.2   Technical Dimensions in Vehicle Electricity Transactions Services

6.2.1   System Model

The VET mechanism consists of three entities: user, registration authority (RA), and blockchain network.

- **Users**. The users include EVs and charging facilities. They can be either payers or payees. Each user is a blockchain client and is able to maintain the blockchain.
- **Registration Authority**. The RA is responsible for the account registration and payment record auditing who is a blockchain client and is able to maintain the blockchain. It also maintains a certificate pool storing all registered accounts.
- **Blockchain Network**. A blockchain network is the blockchain infrastructure consisting of the communication and consensus mechanisms.

In V2G networks, there are many bidirectional interactions between EVs and smart grids regarding electricity transmissions and payment bills. These payment bills are uploaded to a payment platform which completes payments and shares the records. By observing the payment records, users can provide services for other users. For example, EV owners may ask for suggestions on choosing cost-effective time to charge EVs. The electricity payment process in V2G networks has three steps as depicted in Fig. 6.2.

- **Bill Generation**. EVs are granted an access to V2G networks by parking lot-designated facilities and EVs provide services for smart grids through local aggregators (LAs) [214]. EVs have three main states: charging, discharging, and distributed discharging. An EV initiates charging and the electricity is transferred from smart grids to the EV. After charging, the EV has to pay for an electric bill to the smart grid. A smart grid initiates discharging when the grid is overloaded. EVs can assist in alleviating power shortages by discharging to the smart grid. After discharging, EVs can receive financial rewards from the smart. An EV initiates distributed discharging when its battery capacity is lower than a threshold.
- **Payment Execution**. A payment platform executes payment when receiving electricity bills and then shares payment records to users. The payment method has to offer a reliable transaction to guarantee that the transaction is authentic. Meanwhile, the payment method has to support auditability to solve payment disputes.
- **Payment Record Sharing**. Payment records contain transaction information and account information. The transaction information contains total price and unit price. The account information contains the identities of traders. EVs and LAs can provide services based on obtained transaction information. In order to protect user privacy, account information is anonymized. Since a privileged user, i.e., registration authority (RA), is responsible for auditing transactions

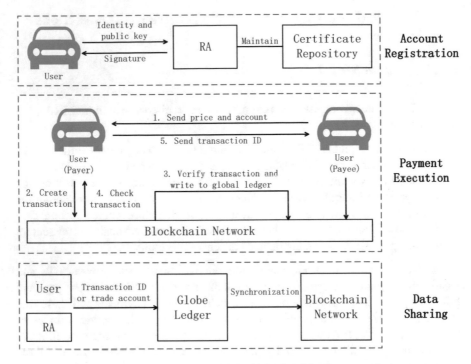

Fig. 6.2 Architecture of the blockchain-enabled vehicle electricity payment scheme

and identifying malicious users, it has to know the transaction and account information. Hence, the payment method has to protect user privacy and enable data auditing.

6.2.2 Threat Model

We mainly focus on two types of threats: privacy disclosure and unreliable payments.

Privacy Disclosure EVs are connected to LAs through insecure wireless connections such that the payment information can be eavesdropped by adversaries. Since payment records are publicly shared, potential adversaries may probe into the sensitive information of EVs. Moreover, an adversary can further acquire user privacy based on payment records by colluding with malicious users.

Unreliable Payments An adversary can try to create unreliable payments, such as fake payments and double-spending. Defending such an attack in a distributed system without a trusted third party is a challenging task.

6.2.3   Design Objectives

We aim to design a vehicle electricity transaction payment mechanism for V2G networks achieving data sharing and privacy protection.

- **Privacy Preservation**. Payment records should be securely and efficiently shared. An adversary cannot infer the real identities of traders from payment records.
- **Effective Auditing**. Even though the payment records are anonymous, an authorized auditor can reveal real identities of traders.
- **Reliable and Efficient Payments**. The payment method should support reliable and efficient payments, i.e., resisting unreliable payments. Moreover, the payment method should have better scalability so as to meet practical transaction requirements in V2G networks.

6.3   Solution

This section presents a solution of blockchain-assisted vehicle electricity transaction services, including account registration, payment execution, and data sharing.

Account Registration   In account registration phases, each user submits an account to the RA for registration and only the signed account are legal accounts in the following electricity transactions.

The user first generates keys (P_k_user; S_k_user), and then sends P_k_user and its identity to the RA. After validating the identity, RA returns to the user a signature δ = Sign(S_k_RA;P_k_user) generated under its private key S_k_RA. RA also stores P_k_user and the identity in the certificate pool. The legal account format is account = (P_k_user, δ). Each user can use P_k_RA to verify the account $v = Verify(P_k\_RA; account.P_k\_user; account.\delta)$.

Users can apply for signatures from the RA for multiple public keys at once and reapply when these accounts are exhausted.

Payment Execution   The payee sends an account information and a unit price to the payer. The payer computes an electricity bill, generates a transaction, and sends the transaction to the blockchain network. After receiving the transaction, the blockchain network verifies the transaction and records the transaction in the blockchain. The payment is completed when the transaction is recorded in the blockchain.

The payer keeps monitoring the blockchain until the payment is completed. Then, the payer sends the transaction ID to the payee. When there is a transaction with the corresponding account and amount information, the payee provides services to the payer, like electricity transmission.

Txid	Txhash								
		Source Field			Price Field		Destination Field		
sn_sf	pre_txhash		pre_sn_df	signature	unit price	sn_df	account		amount
sn_sf	pre_txhash		pre_sn_df	signature	total amount	sn_df	account		amount
		...						...	

Fig. 6.3 Transaction structure for the payments in V2G networks

Data Sharing Since the blockchain is publicly accessible, anyone who needs to view a payment record can download the blockchain with a blockchain client.

EVs and malicious users cannot link the payment record to a specific user because the account information in payment records are pseudonyms. The RA can reveal all the account identities from the certificate pool, effectively auditing payment records.

Transaction and Verification A novel type of transaction structure and a corresponding verification method based on Hyperledger are designed to guarantee the reliability of payment transactions as illustrated in Fig. 6.3. Each transaction has a unique ID $txid$ and a unique hash value $txhash$ as an index of the transaction. The content of a transaction consists of three parts: the source field, price field, and destination field.

A verification method is proposed to detect false transactions to guarantee the payment reliability:

- Verify whether the transaction structures meet the format requirements.
- Verify whether all accounts are legal, including input and output accounts.
- Verify whether all signatures are valid.
- Verify whether the input amount is not less than the total output amount.

6.4 Implementation Scenario

A prototype of the VET mechanism is built on Hyperledger where each user has a Laptop with an Inter 2.5 GHz 64-bit CPU, 8G RAM.

We evaluate the time costs of cryptographic operations using the Elliptic Curve Digital Signature Algorithm on the 256-bit curve secp256k1. The average time costs for generating a new key pair and signing a public key are approximately 3.16 ms and 3.45 ms, respectively.

Figure 6.4 shows the computational costs in the registration process. The computational costs of two schemes increase linearly with the number of accounts. We require that each user chooses ten new public keys every day and conducts the registration process every month. The frequency of such doing approximates one of an individual EV in V2G networks. Therefore, each registration process contains 300 key pair generations and 300 signatures, which consumes about 1.9 s.

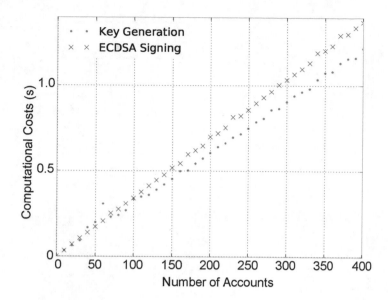

Fig. 6.4 Computational costs in the registration phase

6.5   Summary

This chapter presented a solution of blockchain-enabled vehicle electricity transaction mechanism for V2G networks. Through a registration phase, a new payment method enabled payment auditing while preserving data privacy. Service providers can obtain payment records for data analysis without knowing the real identities or sensitive information of EVs. The blockchain used in the payment method guaranteed the reliability of the payments. When a transaction is recorded on blockchain, the pertinent payment record is tamper-proof. We implemented a prototype of the mechanism and demonstrated its feasibility and effectiveness.

Exercises

6.1 What is the system model for VET services? Refer to Sect. 6.2.1.

6.2 What is the security model in the VET system? Refer to Sect. 6.2.2.

6.3 What are the design goals in the VET system? Refer to Sect. 6.2.3.

6.4 What are the main phases of the VET system? Refer to Sect. 6.3.

6.5 Say Alice is driving an electricity vehicle and she is also transacting electricity to other vehicles through a blockchain, what data will she put on the blockchain?

Chapter 7
Blockchain-Enabled Carpooling Services

7.1 Overview

This chapter aims to introduce a blockchain-enabled technology that is implemented for preserving privacy in carpooling services. Throughout this chapter, mechanisms, models, key terms, and major techniques will be covered in order to provide learners with a holistic view about the application. A privacy-preserving carpooling services (PCS) is a service manner that offers assistance services for carpooling functionality while considering the objective of the privacy protection during the service deliveries.

Carpooling [218, 219] is a common activity of using additional delivery resource for earning benefits [220], e.g., financial earnings or time saving, which frequently takes place in individuals' daily lives. As we know, it will be a good option for many people who need to travel to a certain place when other public/private transportations are unavailable [221, 222]. Knowing others' destinations is a fundamental condition for initializing a carpooling service, because people in a carpool generally have near destinations or on the same route. Hence, each passenger needs to search for a driver who is willing to give a ride for him/her to the desired location.

While people are having the benefit of carpooling, one of the concerns remains to be protecting individuals' privacy as carpooling people may not know each other in most situations. Passengers and drivers all have their privacy concerns [115, 179] before carpooling. These privacy concerns include identity, pickup location, and destination which are contents of carpooling requests and carpooling responses. But encrypting the carpooling requests and carpooling responses will result in an obstacle to carpooling services, since it will be difficult for the service provider to match passengers with drivers. Hence, PCS can be very useful in an application scenario where both passengers and driver want to engage in a carpooling process but they do not want to reveal too much personal information. Meanwhile, sending carpooling requests and carpooling responses to a remote service provider [119, 177] will consume a huge amount of network bandwidth

© Springer Nature Switzerland AG 2019
L. Zhu et al., *Blockchain Technology in Internet of Things*,
https://doi.org/10.1007/978-3-030-21766-2_7

and cause increased response delay. Recently, fog computing [113] was put forth to extend the cloud computing's computing and communicating capabilities to the network edge. It locally pre-processes users' data by fog nodes to improve network performance.

This chapter introduces a solution of the PCS protocol FICA. Specifically, we adopt a private proximity test with location tags [223] to achieve proximity matching and establish a secret key between a passenger and driver; we achieve destination matching by a range query technique [224] efficiently. We also build a private blockchain [25, 225, 226] construct by road-side units (RSUs) to store carpooling records to guarantee data auditability [227]. We put the encrypted carpooling data on the service provider end and store its hash value on the blockchain to reduce storage costs. From the experimental result, we can see that the computational costs caused by the introduction of our blockchain in an acceptable range since the blockchain provides a capability of auditing carpooling records.

The remainder of this chapter is organized as follows: in Sect. 7.2, we present the technical dimensions in privacy-preserving carpooling services. We introduce the basic techniques of privacy-preserving carpooling services in Sect. 7.3.We present a solution of blockchain-assisted privacy-preserving carpooling services in Sect. 7.4 and provide the use case in Sect. 7.5. Lastly, we draw our summary in Sect. 7.6.

7.2   Technical Dimensions in Carpooling Services

This section presents the technical dimensions of privacy-preserving carpooling services (PCS).

7.2.1   Essential Components in Carpooling Services

The PCS system model includes driver, passenger, RSU, cloud server, and trusted authority which are displayed in Fig. 7.1. In this chapter, we use "user" to stand for both passenger and driver.

- **Passenger**: sends a carpooling request to a local RSU and waits for a carpooling response. The carpooling response is returned to from a driver and other passengers who share the driver's vehicle. The passenger's carpooling request contains a pseudo identity, two timestamps, an embedding key, a location hash value, an encrypted drop-off location.
- **Driver**: waits for carpooling requests broadcasted by local RSUs. If he sees a matching request (i.e., the current location, the destination, the condition are

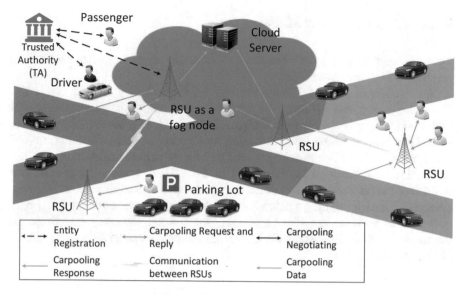

Fig. 7.1 Architecture of fog computing

matched and there are available seats) and he is willing to provide a ride to the corresponding passenger, he will send an encrypted carpooling response to his local RSU. The driver's carpooling response contains a pseudo identity, an encrypted identity, a set of location proofs, a location hash value, and a set of encrypted drop-off locations.

- **RSU**: is a fog node with computation and communication capabilities. An RSU collects passenger's carpooling queries and drivers' carpooling responses in a real-time manner. Then it authenticates users as well as their locations, verifies the integrity of their data, and matches passengers with drivers. It also uploads encrypted carpooling data to the cloud server and maintains a private blockchain with other RSUs. We assume that RSUs are connected to each other through one-hop or multi-hop.

- **Cloud Server**: is a powerful server which collects carpooling queries and carpooling responses from all the RSUs. It answers drivers' traffic queries if asked by RSUs and helps monitor traffic conditions by analyzing vehicles' numbers and trajectories.

- **Trusted Authority** (TA): initializes the privacy-preserving carpooling system, and generates system parameters and keys for the other entities. TA remains offline for most of the time until a user complains about another user. Only TA preserves the ability to disclose the real identify of a targeted user. This centralized role does not conflict with the private blockchain since it is only a system initializer and an identity tracker which stays offline after system initialization.

7.2.2   Cognize Threat Model

The potential security threats in our privacy-preserving carpooling system arise from external and internal adversaries.

We consider that the TA is totally trusted and it is not easy to compromise TA. The cloud server and RSUs are honest-but-curious, which means that they will look into users' privacy such as identity and location [228]. Most of the users are honest and they will send their carpooling data faithfully. Only a small portion of users are malicious. For example, a passenger may perform a location cheating attack, i.e., report a false location [229] to an RSU, to book a driver such that the driver has to drive a long distance to pick up the passenger. A driver may perform a location cheating attack as well. Additionally, a passenger or driver may also conduct criminal activities [230, 231]. An external adversary eavesdrops on the communication channels and launches impersonation attack [232], replay attack [233], and forgery attack.

7.2.3   Anticipated Performance Matrix

We evaluate the performance of the solution of the PCS protocol from two aspects: security and efficiency. First, we will compare the security and privacy properties with existing work. Specifically, we look into user authentication, location authentication, data confidentiality, data integrity, user anonymity, and traceability.

- **User Authentication**. Before a user joins in the carpooling system, his identity has to be authenticated. When an RSU is authenticating the user, the real identity of this user is still kept as a secret from the system authenticator after the identity authentication passes. The illegal users cannot impersonate a legal user so as to participate in the system.
- **Location Authentication**. Before an RSU accepts the location information in a carpooling message of some user, it has to verify whether this user is indeed at the claimed location.
- **Data Confidentiality and Integrity**. The contents of all carpooling messages, including identity, current location, destination, driver condition, and passenger condition, should be preserved from other users, RSUs, and the cloud server. The integrity of the contents in all carpooling messages should be guaranteed, i.e., they are not falsified in transmission by any entity.
- **User Anonymity**. When submitting carpooling requests and responses to the carpooling system, the identities of the users must be protected. It is difficult for any entity to differentiate two carpooling messages, i.e., tell if they come from the same user.
- **Traceability**. The TA reserves the capability to disclose the real identity of a targeted user in case a dispute happens. We claim that the centralized role of TA

is not contradictory to the blockchain since it is only a system initializer and a real identity discloser if needed.

Second, we establish a prototype and examine its running efficiency.

- **Computational cost.** The computing burden of users and RSUs should be as less as possible since the computing power of users' mobile devices and on-board units (OBUs) has limited computing capabilities and the RSUs with constrained powers have to process a number of local data at the same time.
- **Communication overhead**. The communication overhead for users and RSUs should be as low as possible in order to save network bandwidth.

7.3   Basic Techniques of Carpooling Services

This section introduces the basic techniques of privacy-preserving carpooling services.

7.3.1   Anonymous Authentication

An anonymous authentication scheme [159] aims to authenticate a user before the user participates in a system while protecting his real identity from the authenticator. The idea is to assign an anonymous key to the user via a trusted authority, and then the user generates an anonymous certificate as well as a signature as participation qualification. Specifically, it consists of six algorithms:

- Setup(1^k): takes a security parameter 1^k as input and outputs system parameters *params*, two private keys u, v, three public keys $\tilde{A}_1, \tilde{A}_2, \tilde{B}$, and a secure hash function H.
- KeyGen(*params*, i): takes public parameters *params* and a user's identity i as input, and outputs an anonymous key $AK_i = (s_i, S_i)$.
- PseudoGen(*params*, AK): takes public parameters *params* and an anonymous key AK as input, and outputs private keys $(x_1, x_2, \ldots, x_l)$ and corresponding public keys $(Y_1, Y_2, \ldots, Y_l)$.
- Sign(*params*, S, x, Y, m): given *params*, an anonymous key S, a private key x, a public key Y, and a message m, computes an anonymous certificate *Cert* and a signature σ.
- Verify(*params*, $m, \sigma, Cert$): given *params*, m, σ, and *Cert*, verifies the validity of *Cert* and σ. The output is 1 if the verification passes, or $\perp$ otherwise.
- Open(*params*, $Cert$): given *params* and *Cert*, outputs an anonymous key S.

7.3.2 Private Proximity Test with Location Tags

A private proximity test with location tags [223] helps us authenticate a user's location and determine whether two users are in close proximity to each other. Basically, when a user wants to know if he has friends in his proximity, he can send a proximity test request message and location proof with time and location requirements to a server. The server will look for appropriate users. A responding user will also send a location proof to the server and prove that he is at the location he claims to by collecting environment signals to generating location tags according to the request. Then the server, i.e., location authenticator and user matcher, could check if they are close enough by verifying their location proof. Specifically, it contains two steps:

- **Location-Based Handshake.** User Alex sends to the cloud server a request REQ_A including a filtering function φ, a location tag filtering function ϕ, two time-points t_1 and t_2, and a hash function H. When seeing REQ_A, the cloud server searches for candidates based on φ and broadcasts a synchronization message SYN. After receiving SYN, each candidate collects some environment signals and generates a location tag. Alex inserts observations into a Bloom filter [234]. Alex computes a RSA secret and a RSA public keys, and embeds the latter into the location tag. Alex sends the embedded message EMB_A to the cloud server. The cloud server broadcasts EMB_{Alex}. A candidate Betty in Alex's proximity can recover pk_A correctly.
- **Private Proximity Test.** Alex defines a vicinity region and inserts each grid number into a Bloom filter and sends a proximity test request PRO_{Alex} to the cloud server. Betty encrypts her identity with pk_A to form eid_{Betty}, broadcasts it, and collects eid from others. Then Betty sends a response RSP_{Betty} to the cloud server. The cloud server filters candidates for Alex using Bloom and sends remaining eid to Alex. Alex decrypts the eid to get the matched candidate's identity and retrieves her public key from a trusted certificate authority.

7.3.3 Privacy-Preserving Range Query

A privacy-preserving range query scheme [224] provides a way to check if a number falls into a given range without exposing this very number. Specifically, it includes five steps:

- **Prefix Encoding**: given a real number n with the binary form $b_1b_2 \ldots, b_w$, a data owner computes n's prefix family $PF(n)$; given a range $[lb, rb]$, a data user computes a minimum set of prefixes $\mathscr{MS}([lb, rb])$.
- **Tree Construction**: given d different prefixes families, the data owner builds a balanced binary search tree pbt.

- **Node Randomization**: the data owner and the data user previously share o secrets $k_1, \ldots, k_o$. For each prefix pr in a tree node on pbt, the data owner computes $HMAC(k_1, pr), \ldots, HMAC(k_o, pr)$, and computes $HMAC(r, HMAC(k_1, pr)), \ldots, HMAC(r, HMAC(k_o, pr))$, where r is a random number, and inserts them into a Bloom filter. Lastly, the data owner sends pbt as well as encrypted data to a data center.
- **Trapdoor Computation**: given a query range $[lb, rb]$ with z prefixes $pr_1, \ldots, pr_z$ in $MS([lb, rb])$, the data user computes t hashes for each prefix. The trapdoor for $[lb, rb]$ is a matrix $M_{[lb,rb]}$ of $z \cdot t$ hashes. Lastly, the data user sends $M_{[lb,rb]}$ to the data center.
- **Query Process**: given a range query trapdoor $M_{[lb,rb]}$, the data center searches the pbt and checks whether there is a row j in $M_{[lb,rb]}$ such that the result of querying $(r, HMAC(k_i, pr_j))$ in the Bloom filter is 1.

7.3.4 Fog Computing

For the past 6 years, fog computing [113, 235] has been advancing since CISCO first proposed this concept in 2011 [113], which allows billions of connected devices and users in IoT to achieve local communication and local computation at the network edge. Fog computing can reduce the number of data packets transmitted between users and a remote cloud server since users do not have to communicate with the cloud server when there is a local fog node which is considered as a mini-server. Meanwhile, the original communication overhead will decrease the service quality and user experience. It is worth noticing that the utilization of fog nodes can assist the cloud server in completing some computation tasks for local user to a great extent.

The architecture of fog computing is shown in Fig. 7.2. We can observe that the cloud is still in a centralized entity and it is surrounded by several fogs. Each fog is responsible for collecting data from users in its coverage area. After processing user data, each fog returns corresponding results to users and cloud.

7.4 Solution

This section presents a solution of blockchain-assisted privacy-preserving carpooling services FICA in vehicular networks, which mainly has six phases: system initialization, entity registration, carpooling requesting, carpooling responding, carpooling termination and cancellation, and user tracking.

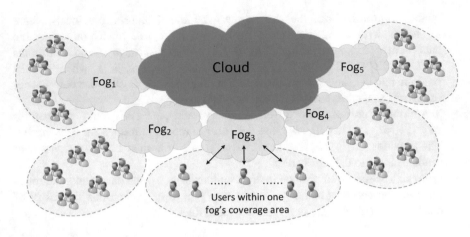

Fig. 7.2 Architecture of fog computing

7.4.1 System Initialization

TA generates three cyclic groups $\mathbb{G}_1$, $\mathbb{G}_2$, $\mathbb{G}_T$ with a prime order q, generators g_1, g_2 for $\mathbb{G}_1$, $\mathbb{G}_2$, and a bilinear map $e : \mathbb{G}_1 \times \mathbb{G}_2 \to \mathbb{G}_T$. TA randomly chooses $a, b \in \mathbb{Z}_q^*$ as private keys, selects a hash function $H : \{0, 1\}^* \to Z_q^*$, and computes $\tilde{A}_1 = g_1^a$, $\tilde{A}_2 = g_2^a$, $\tilde{B} = g_1^b$.

Then, TA chooses a signal filtering function ϕ, a length f_1 and a set of hash functions $\mathscr{H}$, and a length l. TA divides the whole service region into grids $g(i)$s and organizes them in a tree.

Next, TA selects a length w for grid numbers and a length f_2, and a secure keyed hash function $HMAC$.

Finally, TA publishes all the system public parameters $(q, \mathbb{G}_1, \mathbb{G}_2, \mathbb{G}_T, \tilde{A}_1, \tilde{A}_2, \tilde{B}, w, f_1, f_2, l, \mathbf{g}(i), \phi, H, \mathscr{H}, HMAC)$.

7.4.2 Entity Registration

Now a passenger with a real identity P_i participates the carpooling system, TA randomly chooses a number $s_i \in \mathbb{Z}_q^*$ and numbers $k_1, k_2, \ldots, k_o$, and computes $S_i = g_1^{1/(s_i+a)}$. TA stores (P_i, S_i^a) in an identity list and sends back secret keys $k_1, k_2, \ldots, k_o$, anonymous key $AK_i = (s_i, S_i)$. P_i randomly chooses l_i numbers $(x_{i1}, x_{i2}, \ldots, x_{il_i})$ as private keys and computes public keys $Y_j = g_1^{x_j}$ ($j = 1, 2, \ldots, l_i$). In a similar way, a driver D_j receives $k_1, k_2, \ldots, k_o$, AK_j, rp_j, and $\{x_{io}, Y_{io}\}_{o=1}^{l_j}$. An RSU $ID_{\hat{R}}$ obtains from TA a secret key $x_{ID_{\hat{R}}}$ and a public key $Y_{ID_{\hat{R}}}$.

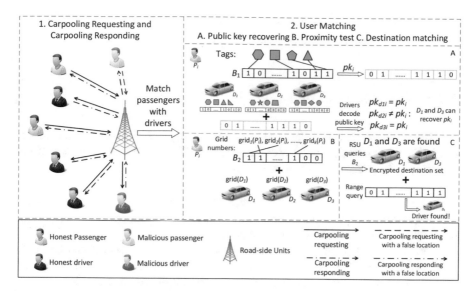

Fig. 7.3 Carpooling requesting, carpooling responding, and user matching

The main phases of FICA, i.e., carpooling requesting, carpooling responding, and user matching, are shown in Fig. 7.3.

7.4.3 Carpooling Requesting

A passenger with a pseudo name pid_{P_i}, two time-points t_1, t_2, a current location loc_i, a drop-off location go_i, a secret key x_i, and a public key Y_i first generates a carpooling request REQ_i:

- P_i collects some environment signals $\mathscr{X}_i(t_1, t_2)$ and generates a location tag $\mathscr{Y}_i(t_1, t_2) = \phi(\mathscr{X}_i(t_1, t_2))$. P_i hashes the observations $y_i(t_1, t_2)$ into a Bloom filter $\{0\}^{f_1}$ through $\mathscr{H}$ to get a location tag $\hat{B}_{i1} = \mathsf{Insert}(H(y_i(t_1, t_2)), \hat{B}_{i1})$. P_i changes loc_i into $\mathbf{g}(P_i)$ to be a set of grids standing for the vicinity region and hashes each grid number into a Bloom filter $\hat{B}_{i2} = \mathsf{Insert}(H(\mathbf{g}_i(P_i)\|pk_i), \hat{B}_{i2})$.
- P_i computes keys $sk_i, pk_i \in \{0, 1\}^l$ and embeds pk_i into $\hat{B}_{i1}$ by calculating: $en_i = \mathsf{Encode}(f_1, l, pk_i)$ and $sh_i = en - \hat{B}_{i1}$ [236].
- P_i chooses go_i with grid number $g(i)$ and changes $g(i)$ to binary form to generate a prefix family $\mathscr{PF}(g(i))$ [224]. P_i randomly chooses a number r_i and computes $\{HMAC(r_i, HMAC(k_j, pf_i))\}_{j=1}^{o}$ for each pf_i. P_i inserts each of them into a Bloom filter $\{0\}^{f_2}$ to get $\hat{B}_{i3}$. Finally, P_i has created a carpooling request

$$REQ_i = \{pid_{P_i}, t_1, t_2, sh_i, \hat{B}_{i2}, \hat{B}_{i3}, r_i\}. \tag{7.1}$$

Next, P_i computes an anonymous certificate $Cert_i$ and a signature σ_i [159], and sends $\{REQ_i, Cert_i, \sigma_i\}$ to a local RSU $ID_{\hat{R}}$.

7.4.4 Carpooling Responding

After receiving an encrypted carpooling request $\{pid_{P_i}, REQ_i, Cert_i, \sigma_i\}$, the local RSU $ID_{\hat{R}}$ first checks the validity of $Cert_i$ and σ_i. If either of them fails, it discards the query or broadcasts a synchronization message:

$$SYN_i = \{t_1, t_2, sh_i\}_{ID_{\hat{R}}} \tag{7.2}$$

to all drivers within its own coverage region.

When seeing SYN_i, a driver D_j (who is willing to pick up this passenger) with a pseudo identity pid_{D_j}, a current location loc_j, a set of drop-off locations $\mathbf{go}_j$, a secret key x_j, and a public key Y_j generates a carpooling response RES_j:

- D_j computes two Bloom filters $\hat{B}_{j1}$ and $\hat{B}_{j2}$. D_j calculates $en_j = sh_i - \hat{B}_{j1}$ and decodes en_j by computing $pk'_i = \mathsf{Decode}(f_1, l, en_j)$.
- D_j chooses a symmetric communication key sk_j^{com}, computes $eid_j = \mathsf{ENC}(pk'_i, pid_{D_j}||sk_j^{com})$, and broadcasts eid_j. D_j gathers $eids$ from nearby drivers to form a group of location proofs $\mathscr{P}_j$. D_j changes loc_j to a grid number $g(D_j)$ and computes $H(g(D_j)||pk'_i)$.
- D_j chooses a grid set $\mathbf{g}(D_j)$. D_j transforms $[g_{min}(D_j), g_{max}(D_j)]$ to a set of prefixes consisting of z prefixes pr_i, $1 < i < z$. D_j calculates a set of hashes $\mathscr{HMAC}_j = \{HMAC(r_i, HMAC(k_1, pr_1)), \ldots, HMAC(r_i, HMAC(k_t, pr_z))\}$. D_j creates a carpooling response:

$$RES_j = \{pid_{D_j}, eid_j, \mathscr{P}_j, H(g(D_j)||pk'_i), \mathscr{HMAC}_j\}. \tag{7.3}$$

Finally D_j generates an anonymous certificate $Cert_j$, a signature σ_j, and sends $\{RES_j, \sigma_j, Cert_j\}$ to the local RSU $ID_{\hat{R}}$.

7.4.5 Carpooling Matching and Uploading

When receiving $\{pid_{D_j}, RES_j, Cert_j, \sigma_j\}$ from a driver D_j, the local RSU $ID_{\hat{R}}$ checks the validity of $Cert_j$ and σ_j. If they pass verification, $ID_{\hat{R}}$ will match the driver with passengers:

- **One-to-many proximity matching**: $ID_{\hat{R}}$ hashes $H(g(D_j||pk'_j))$ into $\hat{B}_{i2}$ and checks whether the result is 1. Assume res_i is the set of remaining $eids$ with query result 1 from querying $\hat{B}_{i2}$.

- **Get-off location matching**: $ID_{\hat{R}}$ hashes $\mathcal{HMAC}_j$ into $\hat{B}_{i3}$ and checks the result. Then $ID_{\hat{R}}$ sorts res_i.

Next, $ID_{\hat{R}}$ returns $\{res_i\}_{ID_{\hat{R}}}$ to P_i. P_i decrypts every eid by computing $pid_{D_j}||sk_j^{com} = \mathsf{Dec}(eid, sk_i)$. Finally, P_i can communicate with D_j under the encryption and decryption of the secret key sk_j^{com} for future communications, such as where to pick up, and where to drop-off.

When P_i and a matched driver D_j start a carpooling trip, they send two confirmations $\{ID_{\hat{R}}, pid_{P_i}, pid_{D_j}, eid_j\}_i$ and $\{ID_{\hat{R}}, pid_{D_j}, pid_{P_i}, eid_j\}_j$ to $ID_{\hat{R}}$. $ID_{\hat{R}}$ returns to P_i and D_j an authentication token $tok_{ij} = \mathsf{Sig}(x_{ID_{\hat{R}}}, pid_{P_i}||pid_{D_j}||t_{ij1})$. This token is designed for further communication with a new RSU after the ride is done.

After P_i arrives at the drop-off location, P_i pays D_j with mobile payment or cash. P_i encrypted i, loc_i', go_i, fair pay_{ij}, and t_{ij2} under Y_i to get

$$E_i = \mathsf{Enc}(x_i, P_i||loc_i'||go_i||pay_{ij}||t_{ij2}||r_i') \tag{7.4}$$

and calculates a hash value $H(E_i)$. D_j encrypted $j, loc_{ij}', go_i, pay_{ij}$, and t_{ij2} with Y_j to obtain

$$E_j = \mathsf{Enc}(x_j, D_j||loc_j||go_j||pay_{ij}||t_{ij2}||r_j') \tag{7.5}$$

and calculates a hash value $H(E_j)$. P_i and D_j exchange $H(E_i)$ and $H(E_j)$ so as to generate signatures $\sigma_i' = \mathsf{Sig}(x_i, \hat{R}|| pid_{P_i}|| Y_i|| H(E_i)|| pid_{D_j}|| Y_j|| H(E_j))$ and $\sigma_j' = \mathsf{Sig}(x_j, \hat{R}|| pid_{P_i}|| Y_i|| H(E_i)|| pid_{D_j}|| Y_j|| H(E_j))$. Finally, P_i and D_j send their carpooling records, encrypted carpooling data, and tokens to a different RSU in the drop-off region.

When receiving a carpooling record

$$\mathcal{R} = \{ID_{\hat{R}}, pid_{P_i}, Y_i, H(E_i), \sigma_i', pid_{D_j}, Y_j, H(E_j), \sigma_j'\}, \tag{7.6}$$

and encrypted data $E_i, E_j, ID_{\hat{R}}$ verifies it and then broadcasts $\mathcal{R}$ if it is valid. We note that Y_i and Y_j can be used to disclose P_i's and D_j's real identities. $ID_{\hat{R}}$ uploads $\{pid_{P_i}, E_i, pid_{D_j}, E_j\}_{ID_{\hat{R}}}$ to the cloud server.

Here, we deem RSUs as stakeholders and they all have their own stakes, i.e., the total number of firstly received carpooling records. The private blockchain $\mathcal{B}$ is constructed as follows:

- Time is parted into a continuous sequence of time slots $\{ts_1, ts_2, \ldots\}$ and in each time slot, a new data block is created and appended to a ledger. RSUs can synchronize with each other and conduct a distributed consensus protocol [75] by the use of a leader selection function $F(\cdot)$. All users are within some RSU's coverage region and they have access to the blockchain via the RSUs.

- A genesis block is B_0 which contains an empty block header, $\{ID_i\}_{i=1}^{R}$, $\{Y_i\}_{i=1}^{R}$, stakes $\{st_i\}_{i=1}^{R}$, and a signature computed by the cloud server. All RSUs records a blockchain $\mathscr{B} = B_0$.
- In each ts_i, every RSU verifies the validity of each received carpooling record $\mathscr{R}$. An RSU L_i is selected to pack a new block and its winning probability p_i is proportional to its current stake. All RSUs run $\mathsf{F}(\cdot)$ which takes RSUs' public keys, stakes, probabilities, and timestamp as inputs, and outputs an RSU $ID_{\hat{W}} \in \{ID_1, ID_2, \ldots, ID_R\}$.
- The winning RSU $ID_{\hat{W}}$ packs a new block B_{ts_i} which includes a block header, RSUs' updated stakes, N carpooling records, and a signature.

$$\sigma_{ID_{\hat{R}}}^{ts_i} = \mathsf{Sig}(x_{\hat{R}}, bn_{ts_i}, H_{ts_i}, HR_{ts_i}, t_{ts_i}, st_{ts_i}). \tag{7.7}$$

Then $ID_{\hat{W}}$ adds this newly packed block to $\mathscr{B}$ and broadcasts it. The blockchain construction is depicted in Fig. 7.4.

Some Discussions

- PoW [237] is not considered in selecting a winning RSU since it requires too much energy consumption and time for miners during conducting extensive hash functions. We choose PoW to reach group consensus because we can use the number of carpooling records as the stake of an RSU.
- The TA is not contradictory to the blockchain design for three reasons. First, the decentration feature of blockchain indicates that the data is stored in a distributed ledger maintained by distributed miners. Second, the task of TA is to register users and RSUs, and then disclose a targeted entity's real identity if needed which is not relevant to the ledger. Third, the TA stays offline after system initialization.

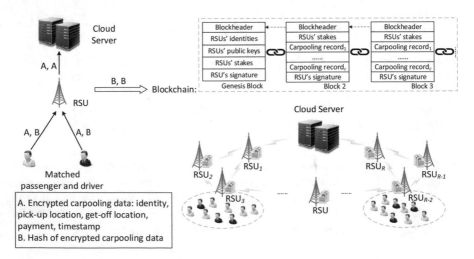

Fig. 7.4 Blockchain construction

- If an RSU is somehow compromised by a malicious adversary and it is controlled to inject meaningless data to the blockchain, we will defend this attack through verifiable computing [238].

7.4.6   Carpooling Termination and Cancellation

After the passenger arrives at the drop-off location, P_i pays the driver by mobile payment or cash. If a user wants to get out of the carpooling requesting or responding process, he can send a cancellation message to the local RSU to remove his request or response from the RSU's processing list.

7.4.7   User Tracking

If a user complains about another user with pid_{P_i} and $Cert_i$ with enough evidence (e.g., video, photo), the TA computes can recover the real identity of driver i through looking up its tracking list. If a matched user is found, TA will punish this targeted user according to the company rules.

7.5   Use Case of Blockchain-Assisted Carpooling in Vehicular Networks

In this section, we analyze the performance of FICA in terms of computational costs and communication overhead. We use Miracl [239] as our cryptographic toolset and an elliptic curve is set as $y^2 = x^3 + 1$ over $\mathbb{F}_p$ with $|p| = 512$ bits. Specifically, the number of passengers and drivers are both within the range [100, 1000], hash function H is SHA256, and the length of pseudo name is 10 bits.

7.5.1   Computational Cost in Carpooling

We initiate a scenario in which an RSU manages data from at most 1000 passengers and 1000 drivers. Then we count the number of the cryptographic operations of each entity. For example, it costs around 31.4 ms for a passenger to generate a carpooling query and it costs about 32 ms for a driver to generate a carpooling response.

We compare the computational cost of FICA with existing schemes: AMA [240] and three sub-schemes DAP-DAD, DAP-ORD, and DAP-EAD in [228]. The results in Figs. 7.5 and 7.6 show that FICA has better performance than other schemes in carpooling requesting and carpooling responding.

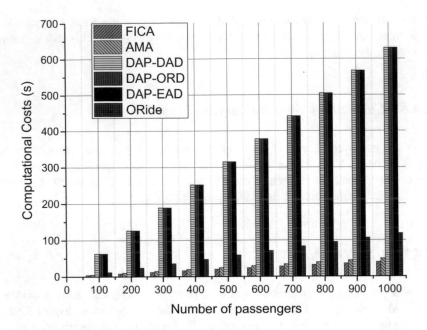

Fig. 7.5 Computational costs in carpooling requesting

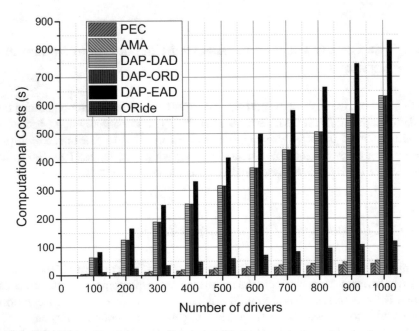

Fig. 7.6 Computational costs in carpooling responding

7.5.2 Communication Overhead in Carpooling

To upload a carpooling request, each passenger has to send $\{REQ, Cert, \sigma\}$ to an RSU with a message length of 0.9 kbytes. To upload a carpooling response, each driver has to send $\{RES, Cert, \sigma\}$ to an RSU with a message length of 1.27 kbytes. The result is shown in Figs. 7.7 and 7.8.

7.5.3 Experiments on Private Blockchain

We also conduct real experiments on the construction of the private blockchain under six different settings. For example, in the first setting, there are ten RSUs and each RSU manages ten passengers and ten drivers, each block after the genesis block has five carpooling records. We set up a thread for each RSU.

The average time cost for each RSU is less than 2 s in carpooling requesting and the average time cost for each RSU is less than 0.4 s in carpooling responding. Generally, the utilization of our blockchain does not incur too much computational burden.

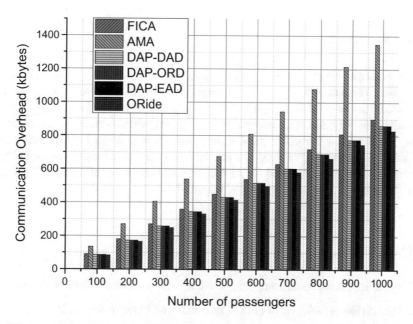

Fig. 7.7 Communication overhead of passengers

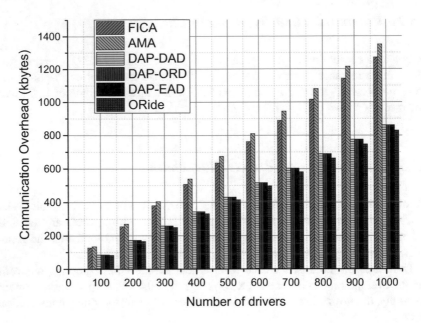

Fig. 7.8 Communication overhead of drivers

7.6 Summary

FICA is secure under a security model where the RSUs and the cloud server are honest-but-curious, and a part of users might be malicious and launch false location attacks. With our delicate design, FICA protects user privacy in a conditional way. Anonymous authentication, private proximity test with location tags, and range query are used to achieve our goals. In addition, we built a blockchain to record all the hashes of carpooling records in a verifiable way and provide data auditability. In the future work, compromised RSUs will be considered. An efficient and privacy-preserving carpooling scheme will be proposed.

Exercises

7.1 What are the essential components in carpooling services? Refer to Sect. 7.2.1.

7.2 What is the security model in the PCS system? Refer to Sect. 7.2.2.

7.3 What are the design goals in the PCS system? Refer to Sect. 7.2.3.

7.4 What are the main phases of the PCS system? Refer to Sect. 7.4.

7.5 How to achieve anonymous authentication? Refer to Sect. 7.4.

7.6 How to achieve private proximity test? Refer to Sect. 7.4.

7.7 How to conduct a privacy-preserving range query? Refer to Sect. 7.4.

7.8 What is the role of fog nodes in the PCS system? Refer to Sect. 7.4.

7.9 Say Alice and Cathy are hailing a ride through a blockchain and Bob is looking for a passenger on the blockchain, what data will they put into the blockchain network?

Chapter 8
Blockchain-Enabled Ride-Hailing Services

8.1 Overview

Ride-hailing services (RHSs) [241, 242] have been developing in the past decade, given the proliferation of sharing economy and the increasing connectivity. RHSs are now serving users (including riders and drivers) worldwide every day [243]. RHSs enable users to upload their ride requests and responses to an RHS platform, and then establish rides through mobile applications. At the same time, edge computing [113, 235, 244] is deployed in vehicular networks to realize real-time services, e.g., monitoring surface conditions [119, 245]. Service providers (SPs), such as Uber [246] and Didi [247], operate on their own datasets [248] which will result in data isolation shown in Fig. 8.1. For example, a Didi[1] rider is now looking for a Didi driver for 20 min, while there are no available Didi drivers nearby. For Didi rider, this situation causes a waste of time and a bad user experience. At this very moment, it is great if an available driver from another SP takes this Didi rider.

This chapter advocates that SPs can collaborate with each other and establish collaborative rides by sharing user data [249]. Generally speaking, users' ride data are handled within their own platforms in the first place and then users have the options to establish a collaborative ride if they are waiting for a long time before being assigned a driver or picking up a rider. The collaborative rides have several benefits: a rider saves waiting time; a driver takes more ride orders; and the two SPs involved in each collaborative ride gain more business profits.

Although there are many benefits from collaborative rides, we face some vital security and privacy problems [115] for the fact that users' privacy (e.g., identity and

---

[1]Didi is a ride-sharing technology conglomerate that is located in Beijing, China. It offers riding services including taxi hailing and private car hailing to millions of users through a smartphone application.

© Springer Nature Switzerland AG 2019   93
L. Zhu et al., *Blockchain Technology in Internet of Things*,
https://doi.org/10.1007/978-3-030-21766-2_8

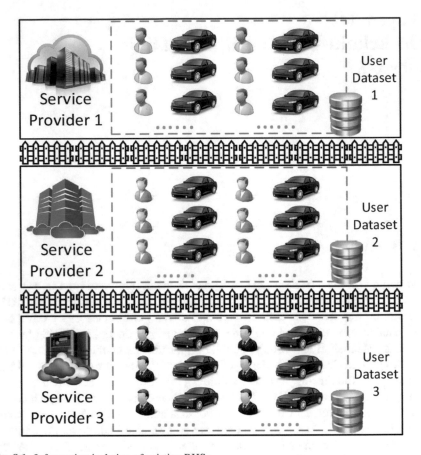

Fig. 8.1 Information isolation of existing RHSs

location) is contained in collaborative rides. First, we have to authenticate the users'
qualifications but keep their real identities anonymous while retaining the ability
of recovering a targeted user' identity. Unlike previous work [158, 159, 250], there
are some unique requirements in collaborative rides: (1) all SPs should be present,
(2) no centralized authority is needed, and (3) the other users remain irrelevant
or anonymous. Second, the users' claimed locations have to be authenticated
to thwart the false location attacks [229]. Third, we have to match riders with
drivers, regarding get-on locations, drop-off locations, and conditions, all based on
ciphertexts. Finally, a rider has to pay for a ride fare when a collaborative ride is
complete. In this work, we resort to an improved Zerocash [251] because the original
design cannot defend the double-spending attack [252].

Since the collaborative rides are established by several SPs, it is important to
write all collaborative rides on a consistent and tamper-resistant ledger [26, 35, 253,
254] so as to guarantee auditability and fairness. A consortium blockchain (*CoB*)

[254] can be used here. All collaborative SPs co-establish the *CoB* and RSUs co-maintain the *CoB*. A *CoB* is a blockchain run by certain parties and it protects transactions among users who do not put their trust fully in others [254]. The *CoB* has been deployed in secure searchable encryption [35] and vehicular networks [26].

Moreover, we also need to consider fairness and efficiency. Fairness guarantees that riders will obtain correct matching results in collaborative ride requesting, drivers will receive corresponding fairs after collaborative rides, and SPs can charge a certain amount of service fees from drivers. Efficiency is another system metric and here we focus on how to match rides with drivers and how to maintain the *CoB* in an efficient way. This chapter introduces CRide: a privacy-preserving collaborative ride-hailing service. Specifically, the private proximity test [223] is used to authenticate users' locations, the privacy-preserving query processing [255] is leveraged to support driver screening and destination matching. A *CoB* among different SPs is constructed to keep track of collaborative rides and deploy RSUs to match riders and drivers and use proof-of-stake (PoS) [75] to achieve group consensus. Last, we build a prototype.

The remainder of this chapter is organized as follows: in Sect. 8.2, we present the technical dimensions in privacy-preserving ride-hailing services. Then, we introduce the basic techniques of privacy-preserving ride-hailing services in Sect. 8.3.We present a solution of blockchain-assisted privacy-preserving ride-hailing services in Sect. 8.4 and provide the use case in Sect. 8.5. Lastly, we draw our summary in Sect. 8.6.

8.2 Technical Dimensions in Ride-Hailing Services

This section presents the technical dimensions of privacy-preserving ride-hailing services (PRHS).

8.2.1 Crucial Components in Ride-Hailing Services

The PRHS system model includes riders, drivers, RSUs, SPs, and a certificate authority CA. Driver, rider, and RSU are denoted by D, R, RS.

CA is responsible for generating keys and choosing a signature scheme and two encryption schemes. It keeps offline after system setup. Each SP is running an RHS platform and it also collaborates with each other to form a collaborative RHS system. SPs create keys for user matching and fare payment. They collect users' ride requests and ride responses through distributed RSUs, construct a consortium blockchain *CoB*, and charge service fees from drivers. By working together, SPs can reveal a targeted user's real identity. Every RSU gathers local ride requests and ride responses, matches riders with drivers as a *CoB* miner, sends encrypted collaborative ride data to SPs, and uploads ride transactions onto the *CoB*. A rider registers to the CA and a SP, puts some money on *CoB*, uploads a ride request to an RSU, and pays a ride fare to a driver. A driver registers to the CA and a SP, responds to a ride request, and receives a ride fare from a rider.

8.2.2   Comprehend Threat Model

The attacks arise from internal and external adversaries.

SPs are not curious about users' privacy and try to complete profiles of users through analyzing collaborative ride data [240]. Also they may collude with RSUs to interrupt the matching procedure. RSUs may collude with SPs or be compromised by adversaries. By collusion attack, it means that the colluding adversaries share information and try to look into the privacy of honest entities. Users are honest-but-curious and a small portion of them can launch false location attacks. External adversaries can eavesdrop on the communication channels and launch impersonation attack, replay attack, and tampering attack.

8.2.3   Expected Performance Objectives

- **Security.** The proposed scheme has to provide data confidentiality, data integrity, anonymous authentication, and location authentication.
- **Privacy.** (1) Anonymity: the identity and location of any user should be protected from SPs, RSUs, and other users during a collaborative ride. (2) Unlinkability: any two requests and responses from the same user should not be linked together. (3) Traceability: any SP should not be able to know the real identity of any user unless all SPs reveal it together. (4) Transaction privacy: the payer, payee, and transferred amount of any transaction should be protected from SPs, RSUs, and other irrelevant users.
- **Auditability.** All the SPs can keep a shared ledger and the collaborative ride transaction in this ledger can be verified by permissioned parties.
- **Fairness.** It guarantees that a rider will be matched with an appropriate driver, a driver will receive a ride fare after a collaborative ride, and the SPs will charge a service fee from their drivers.
- **Efficiency.** Computational costs and communication overhead should be as low as possible during all phases, such as ride requesting, ride responding, and user matching.

8.3   Basic Techniques of Ride-Hailing Services

This section introduces basic techniques of privacy-preserving ride-hailing services.

8.3.1   Conjunctive Query Processing

The privacy-preserving query processing [255] consists of five steps:

- **Index Element Encoding.** For a non-real element, it is encoded into a keyword by attaching its attribute name. For a real element, the prefix family is computed. For instance, given a number a, its prefix family is a set of prefixes $\{a_1 a_2 \ldots a_w, a_1 a_2 \ldots a_{w-1}*, \ldots, a_1 * \ldots *, ** \ldots *\}$. A range $[L, R]$ is transformed into the minimum set of prefixes.
- **Indistinguishable Bloom Filter Construction.** An indistinguishable Bloom filter (IBF) is an array of m twins and hash functions $h_1, h_2, \ldots, h_k, H$. Each twin has two cells where each cell records one bit and the two bits are opposite. The hash function H is used to determine which cell is chosen. For each twin, the value of the chosen cell is 0 and the other cell has 1. Given a keyword a, we hash it into the k twins $B[h_1(a)], B[h_2(a)], \ldots, B[h_k(a)]$ and set the corresponding cells.
- **Indistinguishable Binary Tree Construction.** Given a set of data items $di_1, di_2, \ldots, di_n$ and each data item has a keyword set, an indistinguishable binary tree IBTree is constructed as a balanced full binary tree with each node as an IBF.
- **Trapdoor Computation.** Given a query q, the data user first transforms it into a keyword a by attaching an attribute name. Then he computes $h_1(a), h_2(a), \ldots, h_k(a)$. Next, for each $h_j(a)$, he computes $h_{k+1}(h_j(a))$. The trapdoor for e_i is $\{(h_1(a), h_{k+1}(h_1(a))), \ldots, (h_k(a), h_{k+1}(h_k(a)))\}$.
- **Query Processing.** When receiving a trapdoor td for query q, the cloud server starts the search from the root of the IBTree. For each pair in td, the first hash determines the twin and second hash determines the cell. The search continues along the tree until a leaf node with query hashes all equal to 1.

8.3.2   Anonymous Payment

Zerocash [251] is an anonymous payment scheme and it consists of six steps:

- **System Setup.** A trusted entity runs the setup to generate a set of public parameters pp.
- **Creating Payment Addresses.** Each user generates a new address key pair(s) (add_{pk}, add_{sk}), then it is used to receive coins. add_{pk} is made public, while add_{sk} is kept secret to redeem coins later.

- **Minting Coins.** A user mints a coin by outputting a coin $\mathbf{c}$ and a minting transaction tx^{Dep} and $\mathbf{c}$ has a value v and a coin address add_{pk}.
- **Pouring Coins.** A user pours two coins $\mathbf{c}_3$, $\mathbf{c}_4$ and a pour transaction tx^{Pou} by taking two different coins $\mathbf{c}_1$, $\mathbf{c}_2$ and address secret keys.
- **Verifying Transactions.** Both mint and pour transactions have to be verified by miners before they are accepted.
- **Receiving Coins.** A receiver scans the ledger and retrieves unspent coins to a specific address.

8.4 Solution

This section presents a solution of blockchain-assisted privacy-preserving ride-hailing services CRide, which mainly has six phases: system initialization, entity registration, collaborative ride requesting, collaborative ride responding, collaborative ride termination, and user tracking.

8.4.1 System Initialization

First, CA generates three groups G_1, G_2, G_3 of a prime order p. g_1, g_2 are the generators of G_1, G_2. CA randomly chooses two numbers $u, v \in \mathbb{Z}_p^*$ as private keys, calculates $U_1 = g_1^u$, $U_2 = g_2^u$, $V = g_1^v$ as public keys, and distributes u, v to SPs. Each SP_i has two different shares ss_{i1}, ss_{i2} [256]. CA chooses a hash function $H_1 : \{0, 1\}^* \rightarrow \mathbb{Z}_p^*$ [159].

Second, CA chooses a signature scheme $\mathsf{Sig} := (\mathcal{G}^{\mathsf{Sig}}, \mathcal{K}^{\mathsf{Sig}}, \mathcal{S}^{\mathsf{Sig}}, \mathcal{V}^{\mathsf{Sig}})$, a asymmetric encryption scheme $\mathsf{Enc} := (\mathcal{G}^{\mathsf{Enc}}, \mathcal{K}^{\mathsf{Enc}}, \mathcal{E}^{\mathsf{Enc}}, \mathcal{D}^{\mathsf{Enc}})$, a symmetric encryption scheme $\mathsf{Enc}' := (\mathcal{G}^{\mathsf{Enc}'}, \mathcal{K}^{\mathsf{Enc}'}, \mathcal{E}^{\mathsf{Enc}'}, \mathcal{D}^{\mathsf{Enc}'})$, and generates $\{sk_{SP_i}^{\mathsf{Enc}}, pk_{SP_i}^{\mathsf{Enc}}\}$. CA publishes system parameters $par_1 := (p, G_1, G_2, G_3, g_1, g_2, e, U_1, U_2, V, H_1, par^{\mathsf{Sig}}, par^{\mathsf{Enc}}, \{pk_{SP_i}^{\mathsf{Enc}}\})$ [251].

Next, SPs partition the RHS area into a set $\mathcal{GR} = \{gr[1], gr[2], \ldots, gr[n_G]\}$ and the root grid has number 1. SPs choose par_2 including a filtering function ϕ, a Bloom filter B with length f, a hash function tuple $h := \{h_1, h_2 \ldots, h_o\}$, and $\mathcal{GR}$. SPs choose par_3 including a length w for prefix, an indistinguishable Bloom filter B' with f' twins, a hash function $H'(.) = H'(.)\%2$, and an authentication code HMAC [255].

Last, SPs choose a hash function H_2, three pseudo-random functions $\mathsf{PRF}_x^{add}(i) = \mathsf{PRF}_x(00\|i)$, $\mathsf{PRF}_x^{sn}(i) = \mathsf{PRF}_x(01\|i)$, $\mathsf{PRF}_x^{pk}(i) = \mathsf{PRF}_x(10\|i)$, build C_{POUR} for NP statement POUR, sample a proving key pk_{POUR} and a verifying key vk_{POUR} [251], establish decentralized payment scheme $\Pi^{DAP} := (\mathsf{CreateAdd}, \mathsf{Save}, \mathsf{Pour}, \mathsf{Redeem})$, and generate $SmCo := (\mathsf{Verify}, \mathsf{Hail}, \mathsf{Match})$. SPs set $par_4 := (H_2, \mathsf{PRF}_x^{add}(i), \mathsf{PRF}_x^{sn}(i), \mathsf{PRF}_x^{pk}(i), pk_{\mathsf{POUR}}, vk_{\mathsf{POUR}}, SmCo)$.

8.4.2 Entity Registration

A rider R_i from SP_z participates in the collaborative RHS system by registering to CA. CA randomly chooses a secret key SK_{R_i}, calculates $\{\mathscr{E}\}^{\mathsf{Enc}}(R_i\|SK_{R_i})$ and $\{pk_{SP_j}^{\mathsf{Enc}}\}_{j=1, j\neq z}^{s}$, randomly chooses a number $r_{R_i} \in \mathbb{Z}_p^*$, calculates $\hat{r}_{R_i} = g_1^{1/(r_{R_i}+u)}$, and returns SK_{R_i}, $ak_i := (r_{R_i}, \hat{r}_{R_i})$, $\{\mathscr{E}\}^{\mathsf{Enc}}(R_i\|SK_{R_i})$ as well as a signature $\sigma_{R_i}^{CA}$ to R_i. Then, R_i sends $(\hat{r}_{R_i}, \{\mathscr{E}\}^{\mathsf{Enc}}(R_i\|SK_{R_i}), \sigma_{R_i}^{CA})$ to SP_z and randomly chooses l_i numbers $\{x_{ij}\}_{j=1}^{l_i}$ as private keys, and calculates public keys $\{Y_{ij} = g_1^{x_{ij}}\}_{j=1}^{l_i}$.

Next, R_i randomly selects $K + 1$ secret keys $\mathscr{SK}_{R_i} := \{sk_{R_i1}, sk_{R_i2}, \ldots, sk_{R_iK+1}\}$ and generates a pseudo-random hash function set $h'_{R_i}(.) := \{h'_{R_i1}(.), h'_{R_i2}(), \ldots, h'_{R_iK}(.)\}$, where $h'_{R_ij}(.) := \mathsf{HMAC}_{sk_{R_ij}}(.)\%f'$ and $h'_{R_iK+1}(.) = \mathsf{HMAC}_{sk_{R_iK+1}}(.)$.

Last, R_i calculates $(sk_{R_i}^{\mathsf{Enc}}, pk_{R_i}^{\mathsf{Enc}}) = \mathscr{K}^{\mathsf{Enc}}(par_{\mathsf{Enc}})$, randomly chooses a seed $s_{R_i}^{sk}$, calculates $s_{R_i}^{pk} = \mathsf{PRF}_{s_{R_i}^{sk}}^{add}(0)$, and sets $add_{R_i}^{pk} := (s_{R_i}^{pk}, pk_{R_i}^{\mathsf{Enc}})$, $add_{R_i}^{sk} := (s_{R_i}^{sk}, sk_{R_i}^{\mathsf{Enc}})$. R_i buys several tokens from SP_z as ride fares.

Similarly, a driver D_j registers to the CA and a SP, and obtains $(SK_{D_j}, ak_{D_j}, \{\mathscr{E}\}^{\mathsf{Enc}}(D_j\|SK_{D_j}), \sigma_{D_j}^{CA}, add_{D_j}^{pk}, add_{D_j}^{sk})$.

SPs deploy RSUs by allowing them to participate in the collaborative RHS system and match users. Each RSU RS_m has $add_{RS_m}^{sk} := (s_{RS_m}^{sk}, sk_{RS_m}^{\mathsf{Enc}})$, an address public key $add_{RS_m}^{pk} := (s_{RS_m}^{pk}, pk_{RS_m}^{\mathsf{Enc}})$, and a public signing key pair $(sk_{RS_m}^{\mathsf{Sig}}, pk_{RS_m}^{\mathsf{Sig}})$.

8.4.3 Ride Requesting

A rider R_i from SP_z is hailing a collaborative ride in the coverage area of a local RSU RS_m. First, R_i saves two coins $\mathbf{c}_1, \mathbf{c}_2$ of amount v_1, v_2 at RS_m as a prepaid fare. R_i randomly samples a PRF^{sn} seed τ, two trapdoors tr_1, tr_2, and calculates $cm_1 := \mathsf{Com}_{tr_1}(s_{R_i}^{pk}\|\tau)$, $cm_2 := \mathsf{Com}_{tr_2}(v_1\|cm_1)$, where Com is a commitment scheme [251]. R_i sets $\mathbf{c}_1 := (add_{R_i}^{pk}, v_1, \tau, tr_1, tr_2, cm_2)$ and a save transaction

$$\mathsf{tx}_{R_i1}^{\mathsf{Dep}} := (cm_2, v_1, cm_1, tr_2, time). \tag{8.1}$$

When obtaining $\mathbf{c}_2$, $\mathsf{tx}_{R_i2}^{\mathsf{Dep}}$, R_i uploads $(\mathbf{c}_1, \mathbf{c}_2)$ and $(\mathsf{tx}_{R_i1}^{\mathsf{Dep}}, \mathsf{tx}_{R_i2}^{\mathsf{Dep}})$ via RS_m to coin pool CP and transaction pool TP.

Second, R_i calculates a pseudo-id $pid_{R_i} = H(R_i\|RS_m\|time)$, gathers signals during (t_1, t_2), inserts collected observations $y(t_1, t_2)$ into $B_{R_i1} = \{0\}^f$ to get a location tag $B_{R_i1} := \mathsf{Ins}(h(y_i(t_1, t_2)), B_{R_i1})$ [223]. R_i calculates RSA keys

sk_{R_i}, $pk_{R_i} \in \{0,1\}^{len}$, embeds pk_{R_i} into B_{R_i1} : $En_{R_i} = \mathsf{Encode}(f, len, pk_{R_i})$, and calculates $S_{R_i} = En_{R_i} - B_{R_i1}$. R_i changes the current location loc_{R_i} into $\mathscr{GR}_{R_i}$, and inserts grid numbers into $B_{R_i2} := \mathsf{Ins}(h(\mathscr{GR}_{R_i}||pk_{R_i}), B_{R_i2})$.

Third, R_i encrypts $\mathscr{SK}_{R_i}$ and $r_{B'}$ under pk_{R_i} and gets $E_{R_i} = \mathscr{E}^{\mathsf{Enc'}}(pk_{R_i}, \mathscr{SK}_{R_i}|| r_{B'})$. For conditions which are not real numbers, R_i transforms it to a keyword by and gets a set of condition keywords $\mathscr{W}_{R_i}$ [255]. For each keyword w_j, R_i transforms it into f' twins $B'_{R_i1} := \mathsf{Ins}(h'(w_j), B'_{R_i1})$. For every $B_{R_i1}[h'_i(w_j)]$, R_i sets $B'_{R_i1}[h'_i(w_j)][H'(h'_{K+1}(h'_i(w_j)) \oplus r_{B'})] = 1$ and $B'_{R_i1}[1 - h'_i(w_j)][H'(h'_{K+1}(h'_i(w_j)) \oplus r_{B'})] = 0$. For real grid number gr_{R_i}, R_i calculates its prefix family, transforms it into a keyword, and calculates a similar B'_{R_i2}. Next, R_i builds an IBTree T_{R_i} using B'_{R_i1} and B'_{R_i2}. R_j now obtains an encrypted collaborative ride data packet $Pa_{R_i} = \mathscr{E}^{\mathsf{Enc'}}(SK_{R_i}, loc_{R_i}||\mathscr{W}_{R_i}||gr_{R_i}||time)$.

Last, R_i generates a collaborative ride request

$$Req_{R_i} := (t_1, t_2, S_{R_i}, B_{R_i2}, E_{R_i}, T_{R_i}, Pa_{R_i}), \qquad (8.2)$$

calculates $C_{R_i} := \{Y|| V_1|| V_2|| \tilde{h}|| \tilde{s}_1|| \tilde{s}_2|| \tilde{s}_3\}$ and σ_{R_i} on Req_{R_i} [159], and sends $(pid_{R_i}, Req_{R_i}, C_{R_i}, \sigma_{R_i})$ a local RSU RS_m. The main phases of CRide are depicted in Fig. 8.2.

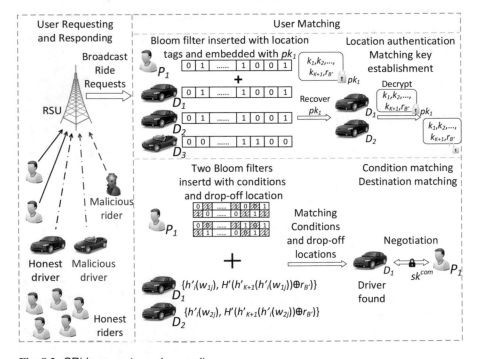

Fig. 8.2 CRide requesting and responding

8.4.4 Ride Responding

When receiving Req_{R_i}, the local RSU RS_m verifies C_{R_i}, σ_{R_i} [159], and cm_2. RS_m sets $cm' := \mathsf{Com}_{tr_2}(v_1||cm_1)$ and accepts $\mathsf{tx}_{R_i 1}^{\mathsf{Dep}}$ if $cm' := cm_2$, otherwise drops it. In a similar way, RS_m verifies $\mathsf{tx}_{R_i 2}^{\mathsf{Dep}}$. RS_m sends $\mathbf{c}_1, \mathbf{c}_2$ to CP and $\mathsf{tx}_{R_i 1}^{\mathsf{Dep}}, \mathsf{tx}_{R_i 2}^{\mathsf{Dep}}$ to TP.

Then, RS_m broadcasts a hailing message $hail_{R_i} := \{t_1, t_2, S_{R_i}, E_{R_i}\}$ to drivers in its coverage area. When receiving $hail_{R_i}$, an available driver D_j from $SP_{z'}$ at loc_{D_j} calculates a pseudo-id pid_{D_j}, calculates a similar B_{j1}, and recovers a pk'_{R_i} from S_{R_i} [223]. D_j chooses a communication key $sk_{D_j}^{com}$, calculates $eid_{D_j} = \mathscr{E}^{\mathsf{Enc}}(pk'_{R_i}, pid_{D_j}||sk_{R_j}^{com})$, and broadcasts eid_{D_j}. D_j gathers eids from nearby drivers to generate a location proof set $\mathscr{P}_{D_j}$, and calculates $h(gr_{D_j}||pk'_{R_i})$.

Next, D_j decrypts E_{R_i} under pk'_{R_i} to get $(\mathscr{SK}_{R_i}||r_{B'}) = \mathscr{D}^{\mathsf{Enc'}}(pk'_{R_i}, E_{R_i})$. Given $\mathscr{W}_{D_j}$, D_j calculates $\mathscr{TR}_{D_j 1}$ including $\{h'(w), H'(h'_{R_i K+1}(h'_{R_i}(w)) \oplus r_{B'})$ for every keyword w. D_j calculates trapdoors $\mathscr{TR}_{D_j 2}$. D_j creates an encrypted data packet $Pa_{D_j} = \mathscr{E}^{\mathsf{Enc'}}(SK_{D_j}, loc_{D_j}||\mathscr{W}_{D_j}||gr_{D_i}||time)$. D_j generates a collaborative ride response

$$Res_j := (eid_{D_j}, \mathscr{P}_{D_j}, h(gr_{D_j}||pk'_{R_i}), \mathscr{TR}_{D_j 1}, \mathscr{TR}_{D_j 2}, Pa_{D_j}), \tag{8.3}$$

and sends $(pid_{D_j}, Res_{D_j}, C_{D_j}, \sigma_{D_j})$ to RS_m.

Last, RS_m hashes $h(gr_{D_j}||pk'_{R_i})$ and $\mathscr{TR}_{D_j 1}, \mathscr{TR}_{D_j 2}$ into $B_{R_i 2}$ and $\mathscr{T}_{R_i}$, sends back to R_i an eid_{D_j} in $\hat{B}_{i2}$ and T_{R_i}. R_i decrypts $pid_{D_j}||sk_{R_j}^{com} = \mathscr{D}^{\mathsf{Enc}}(sk_{R_i}, eid_{D_j})$ and communicates with D_j under sk_j^{com} to negotiate a pickup location and drop-off location. After receiving a confirmation message from R_i and D_j, RS_m sends Pa_{R_i}, Pa_{D_j} to SPs and sends a handshake transaction $\mathsf{tx}_{R_i D_j}^{Han}$ as well as a signature to TP:

$$\mathsf{tx}_{R_i D_j}^{Han} = (pid_{R_i}, pid_{D_j}, RS_m, V_1^{R_i}, V_2^{R_i}, V_1^{D_j}, V_2^{D_j}, H_2(Pa_{R_i}||Pa_{D_j}), time). \tag{8.4}$$

8.4.5 Ride Termination

When the rider arrives at the drop-off location, the rider R_i pays a fare to a matched driver D_j by dividing previously saved two coins $\mathbf{c}_1, \mathbf{c}_2$ to two new coins: the first coin is for a refund and the second coin is for the driver.

Then R_i generates $(sk^{\mathsf{Sig}}, pk^{\mathsf{Sig}})$, calculates $h^{\mathsf{Sig}} = H_2(pk^{\mathsf{Sig}}), h_1 = \mathsf{PRF}_{s_{R_i 1}^{sk}}^{pk}(h^{\mathsf{Sig}}), h_2 = \mathsf{PRF}_{s_{R_i 2}^{sk}}^{pk}(h^{\mathsf{Sig}})$, sets $\overrightarrow{x} := (root, sn_1, sn_2, cm_{12}^{new}, cm_{22}^{new}, 0, h^{\mathsf{Sig}}, h_1, h_2)$, $\overrightarrow{a} := (path_1, path_2, \mathbf{c}_1, \mathbf{c}_2, add_{R_i 1}^{sk}, add_{R_i 2}^{sk}, \mathbf{c}_1^{new}, \mathbf{c}_2^{new})$, calculates a proof π for $\overrightarrow{x}$, sets $M := (\overrightarrow{x}, \pi, \mathbf{C}_1, \mathbf{C}_2)$, calculates $\sigma = \mathscr{S}^{\mathsf{Sig}}(sk^{\mathsf{Sig}}, M)$, sets a pour transaction

$$\mathsf{tx}_{R_i}^{\mathsf{Pou}} := (root, sn_1, sn_2, cm_{1,2}^{\mathsf{new}}, cm_{2,2}^{\mathsf{new}}, 0, info, pk^{\mathsf{Sig}}, h_1, h_2, \pi, \mathbf{C}_1, \mathbf{C}_2, \sigma, time). \tag{8.5}$$

Next, $RS_{m'}$ verifies $\mathsf{tx}_{R_i}^{\mathsf{Pou}}$. If the verification has passed, $RS_{m'}$ sends $\mathbf{c}_1^{\mathsf{new}}$, $\mathbf{c}_2^{\mathsf{new}}$ to CP and $\mathsf{tx}_{R_i}^{\mathsf{Pou}}$ to TP, otherwise drops it.

Last, D_j redeems the ride fare $\mathbf{c}_2^{\mathsf{new}}$ through a SP by computing $(v_2, \tau_2^{\mathsf{new}}, tr_{21}^{\mathsf{new}}, tr_{22}^{\mathsf{new}}) = \mathscr{D}^{\mathsf{Enc}'}(sk_{D_j}^{com}, \mathbf{C}_2)$, verifies whether $cm_{22}^{\mathsf{new}} := \mathsf{Com}_{tr_{22}^{\mathsf{new}}}(v_2^{\mathsf{new}} || \mathsf{Com}_{tr_{21}^{\mathsf{new}}}(s_{D_j}^{pk,\mathsf{new}} || \tau_2^{\mathsf{new}}))$ and $sn_2' = \mathsf{PRF}_{add_{D_j}^{sk}}^{sn}(\tau_2)$ is not on CP. If the verification passes, D_j outputs $\mathbf{c} := (add_{D_j}^{pk}, v_2^{\mathsf{new}}, \tau_2^{\mathsf{new}}, tr_{21}^{\mathsf{new}}, tr_{22}^{\mathsf{new}}, cm_{12}^{\mathsf{new}})$ and pays a service fee to the SP.

Construction of the Consortium Blockchain We use PoS mechanism to periodically elect a leader RSU to generate a new block and avoid forks. Rather than RSUs spending computational resources and time on the leader election process, they instead run a function that randomly selects one RSU proportionally to the stake or balance that each RSU possesses according to the current blockchain ledger. The construction of the consortium blockchain is depicted in Fig. 8.3.

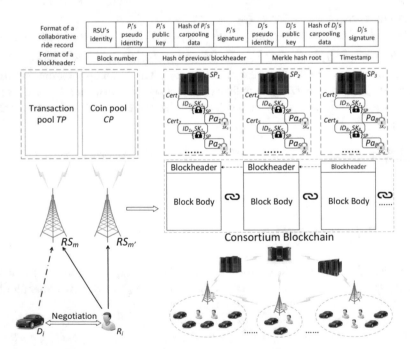

Fig. 8.3 Construction of the consortium blockchain

8.4.6   User Tracking

If a user files a complaint against a targeted user i with pid_i and C_i, SPs recover u, v from secret shares, and calculate $V_2^u/V_1^v = \hat{r}_i^u \cdot V^{uro}/U_1^{vro} = \hat{r}_i^u \cdot g_1^{uvro}/g_1^{uvro} = \hat{r}_i^u$, and recover the encrypted identity of i in their tracking list. If an item is found at SP_z, SP_z first requests other SPs to decrypt $\{\mathcal{E}\}^{\mathsf{Enc}}(i||SK_i)$ to get $i||SK_i = \{\mathcal{D}\}^{\mathsf{Enc}}(\{\mathcal{E}\}^{\mathsf{Enc}}(i||SK_i))$, and recover the collaborative ride data $(time||loc_i||\mathcal{W}_i||gr_i) = \mathcal{D}^{\mathsf{Enc}'}(\mathcal{E}^{\mathsf{Enc}'}(SK_i, Pa_i))$. Finally, SPs add the targeted user in a co-shared blacklist.

8.5   Use Case of Blockchain-Assisted Ride-Hailing in Vehicular Networks

A prototype of CRide is built and then evaluated regarding its computational costs and communication overhead.

8.5.1   Implementation Details

Hundred riders/drivers and 10 RSUs are instantiated on a laptop and 10 desktops. 3 Amazon cloud servers (Intel Broadwell E5-2686v4 8 vCPU@2.3 GHz, 8 GB memory, 64-bit Microsoft Server 2016 Datacenter) [257] are rented. The communication between the laptop and RSU/SP, and between RSUs and SPs uses Socket programming. The lengths of p and q are 160 and 512, respectively. The numbers of grids and transactions (in a block) is 1023 and 8, respectively. The hash functions are SHA256, SHA384, and SHA512. The values of l, f, f', o, K are 10, 1000, 1000, 3, 2.

8.5.2   Computational Costs

In collaborative ride requesting, each rider performs $2Hash$ in Pour, $(3 + 7o)Hash + 2\mathcal{E}^{\mathsf{Enc}'} + 2oXOR + 9Ex_1 + 2Mu_1 + Di_1 + 2BP + Di_3 + 2Hash + 5Add + 4Mul + Div + \mathcal{D}^{\mathsf{Enc}}$ in Request. In collaborative ride responding, each driver executes $(2+5o)Hash + \mathcal{E}^{\mathsf{Enc}} + \mathcal{D}^{\mathsf{Enc}'} + 2oXOR + \mathcal{E}^{\mathsf{Enc}'} + 9Ex_1 + 2Mu_1 + Di_1 + 2BP + Di_3 + 2Hash + 5Add + 4Mul + Div$ in Respond. As shown in Table 8.1, the main time cost for a rider is Request which takes approximately 56 ms, each driver spends 55 ms on responding to a hailing message. A selected RSU spends 1.73 ms on preparing a new block.

Table 8.1 Computational costs

Rider (ms)				Driver (ms)			RSU (ms)
Register	Deposit	Request	Pour	Register	Respond	Redeem	BICr.
37.00	0.06	56.47	0.05	37.00	55.23	0.06	1.73

Table 8.2 Communication overhead

Rider (KB)			Driver (KB)		RSU (KB)
Deposit	Request	Pour	Respond	Redeem	BlockCreation
0.61	0.90	1.02	0.92	0.21	0.26

A rider R_i first sends two coins, two deposit transactions to CP and TP and the bit length is 0.61 kilobyte (KB). Then, R_i uploads 0.90 KB of a pseudo-id, a ride request, an anonymous certificate, and a signature to an RSU. Last, R_i sends 1.02 KB of two new coins and a pour transaction to CP and TP. The detailed communication overhead is shown in Table 8.2.

8.6   Summary

This paper proposed that different RHS platforms could collaborate with each and establish collaborative rides in order to provide more ride-hailing services for user. We proposed a privacy-preserving collaborative ride-hailing service CRide. Specifically, we use private proximity test, privacy-preserving query processing to protect users' privacy and we construct a consortium blockchain among different RHS platforms to record collaborative rides. CRide can protect user privacy in collaborative rides and achieve anonymous payment between riders and drivers. Experimental analysis demonstrates the performance of CRide.

Exercises

8.1 What are the essential components in ride-hailing services? Refer to Sect. 8.2.1.

8.2 What is the security model in the PRHS system? Refer to Sect. 8.2.2.

8.3 What are the design goals in the PRHS system? Refer to Sect. 8.2.3.

8.4 What are the main phases of the PRHS system? Refer to Sect. 8.4.

8.5 How to achieve conjunctive query processing? Refer to Sect. 8.4.

8.6 How to achieve anonymous payment? Refer to Sect. 8.4.

8.7 How to construct a blockchain in the PRHS system? Refer to Sect. 8.4.

8.8 How to track a malicious user in the PRHS system? Refer to Sect. 8.4.

8.9 Say Alice is hailing a ride through a blockchain. Alice is a patient who sees a dentist every month. What is her privacy concerns in ride hailing?

Part IV
Future Research Directions and Discussions

Chapter 9
Exploring Topics in Blockchain-Enabled Internet of Things

9.1 Overview

Given the distinguishing features mentioned earlier, blockchain has taken the world by storm in the last decade [258] for its ability to innovate existing applications, especially IoT application. A blockchain-enabled IoT application can establish a distributed and tamper-resistant ledger, audit previous transactions, and provide partial anonymity. Although blockchain has been identified as a key platform-enabling technology and it is embracing a peak of development, more research efforts must be put into building future blockchain-enabled IoT application systems.

First, a further integration of blockchain with existing IoT systems is on the way. Despite the fact that blockchain has already shown great influences after merging with IoT systems, there are still potentials for blockchain in (possibly new) IoT systems. Next, managing trust roots in blockchain's consensus mechanism. It is twofold in IoT scenarios. The first one is how to trust other users or at least consider them as rational players in the system. The other one is trust of data and it refers to the authenticity of each uploaded data from various users. Then, providing efficient and high-capacity data storage remains a challenging issue since blockchain only has a limited storage space and resorting to an external cloud server will incur new efficiency and security issues. Furthermore, enabling data analysis is a promising feature for blockchain-enabled IoT with amazing advances in machine learning [259–266] and data science [267] in an era of big data [268–274]. There is a huge opportunity to leverage cognitive capabilities into blockchain-enabled IoT systems in order to make them more intelligent. Last but not the least enhancing data security and user privacy is another difficult task. Classic security protection is modern, but cryptographic primitives may be too complicated and time-consuming for blockchain. A new privacy standard, i.e., differential privacy [175] can be considered here.

© Springer Nature Switzerland AG 2019
L. Zhu et al., *Blockchain Technology in Internet of Things*,
https://doi.org/10.1007/978-3-030-21766-2_9

Therefore, this chapter discusses several research topics in blockchain-enabled IoT systems: (1) a further integration; (2) managing trust; (3) providing efficient and high-capacity data storage; (4) enabling data analysis; and (5) enhancing data security and user privacy.

9.2   Future Research Directions

9.2.1   A Further Integration

As introduced in Chap. 2, blockchain is classified into public blockchain, consortium blockchain, and private blockchain. Previous works are mostly pioneering and tentative efforts on integrating blockchain with IoT systems [299–301]. Some potential research directions are given as follows.

Public Blockchain Public blockchain originates from Bitcoin and its main application scenario is finance. The adoption of existing public blockchains is still limited since some of them has a low transaction capacity, e.g., Bitcoin. Although other alternatives show a promising improvement, e.g., Ethereum, many miners in these blockchains are merely participating for valuable coin rewards. Therefore, public blockchain designers should focus on blockchain functionalities added to IoT systems, instead of turning blockchain into a dispensable accessory.

Consortium Blockchain Business partners establish a consortium blockchain to share business transaction and deals in order to reduce data inconsistency and improve operation efficiency. But the application scenarios are limited in big cooperations or banks. This is because constructing such a consortium blockchain requires a lot of resources, e.g., equipment investment, personnel training [275], and maintenance cost. In addition, when it comes to data sharing in business, some commercial companies and sensitive departments may naturally refuse to do it in the first place since some business secrets may be leaked during data sharing. Hence, how to appropriately establish a consortium blockchain for such groups is worthy studying.

Private Blockchain A private blockchain is built by one entity that is similar to a centralized server. It is more suitable for sensitive applications, such as military. An obvious problem that a consortium blockchain may suffer from is the reliability of the centralized entity. If it is compromised, the whole blockchain will be in chaos. A migration of some responsibilities from the centralized entity to downstream entities is a possible solution to this problem. Meanwhile, some efficiency may be sacrificed during the migration.

9.2.2 Managing Trust

Trust [276, 277] is an intrinsic issue in blockchain and it focuses on how to reach an agreement in a decentralized network. Several consensus mechanisms are proposed to solve this problem. However, it does not address the trust towards miner (i.e., miner authentication) or the trust towards data (i.e., data utility).

Trust Towards Miner The first trust issue only exists in public blockchain, since consortium and private blockchains have a stringent requirement for participation approval. In public blockchain, anyone can register to the blockchain network without using any identifiable information and become a legal miner. Since many blockchain applications are enforced with additional functions to provide services, e.g., Permacoin [278] and Filecoin [279], users are concerned about who are performing the underlying computation and whether they are trustworthy?

A simple solution is to keep records of all miners or leverage conditional privacy [280] to anonymously authenticate miners, but this is utterly contradictory to the decentralized characteristics of blockchain. One way of easing the above concern is to divert attention from miners to their data: as long as their data are usable and reliable, people may not dwell on the trust towards miner. Based on this assumption, mechanisms aiming at keep the miners rational can be designed. Here, being rational signifies that miners will follow blockchain rules and work towards a common good. Besides, penalty is an auxiliary method for misbehavior.

Trust Towards Data The second trust issue refers to the authenticity and utility of each uploaded data from various users. One common problem of blockchain is the transactions on blockchain are only a proof of existence with integrity checked, and they do not guarantee that the information contained in the transactions is authentic or useful [280].

Town Crier [281] strengthens the importance of trustworthy data feeds by proposing an authenticated data feed system which bridges a blockchain network with a trusted hardware to retrieve source-authenticated data from HTTPS-enabled websites to relying smart contracts. Unfortunately, this kind of approach is only effective if the server is trusted and the data from the server is authentic, which do not solve the problem fundamentally bottom.

A feasible solution is to introduce truth discovery [282] (TD) into blockchain-enabled IoT systems. Truth discovery is about seeking reliable information and distilling the truth from a set of noisy data collected from different data reporters. When combining TD with blockchain, we can iteratively assign different weights to miners based on the quality of their computation results, which is analogous to stakes, and then calculate a truth from the weighted average of all miners' data until the truth converges.

9.2.3   Providing Efficient and High-Capacity Data Storage

Originally, blockchain is not designed for storage digital files, but business solutions favor data sharing to a great degree. These digital files are of large size and various types, and their amount is growing exponentially over time. Since most blockchain systems utilize a key-value data model with limited storage space, bandwidth, and transaction throughput, novel methods and approaches are called for supporting multiple types of data and large digital files.

First of all, it is only a matter of time for blockchain to support multiple types of data to store as a further integration of blockchain with IoT systems continues. Because data types vary with different IoT systems, this will urge blockchain to achieve a high level of compatibility during integration.

Intuitively, it is infeasible to store large files directly on blockchain. One possible solution [258] is to first store a large file in a cloud server and then put its hash and index on blockchain for quick verification and retrieval. However, this solution requires a compact combination of blockchain and cloud server. Another solution is to data slicing, i.e., slit a large file into pieces and store them on blockchain. This solution has two technical aspects: how to slice a file in order to restore it later and how to store it in a distributed blockchain network while guaranteeing storage efficiency.

9.2.4   Enabling Data Analysis

Blockchain is basically a transaction pool and the transactions are mainly related to financial activities [283]. If data analysis [16, 284–288] is enabled on blockchain, more internal patterns [289] can be acquired in IoT systems, such as electricity consumption trend estimation in smart grid, road condition monitoring in vehicular networks, and accident prevention in industrial networks.

One way to achieve this goal is to make use of a machine learning algorithm [290] or a processing system (e.g., MapReduce [291]) running on blockchain data directly. Users can implement programs to browse blockchain data and obtain a result. If data analysis is integrated with mining process at miner end, it will be more intriguing to see what blockchain can be transformed into. On the other hand, blockchain data are usually encrypted or perturbed to protect confidentiality, thereby presenting a challenging task for data analysis upon these processed data.

Apart from the abovementioned patterns, there are also some sensitive IoT systems, such as military and food supply chain. Data analysis would be even more critical to these blockchain-enabled IoT systems.

9.2.5   *Enhancing Data Security and User Privacy*

Blockchain-enabled IoT systems involve many confidential and sensitive data, such as financial deals, treatment records, smart meter readings, which needs security and privacy protection mechanisms.

Security   Other than focusing on confidentiality, integrity, and authentication, the study of access control [292] in blockchain is well worth efforts since many IoT systems address the data access control problem. A user submits an access request to an authorization entity in the form of a GetAccess transaction [293] and the entity broadcasts the transaction and checks it according to access policies. If granted, the transaction and the access history will be recorded on blockchain and the user can access desired resources. Here, a consortium blockchain or private blockchain can remove the verification part since all miners are mutually trusted and invited into the blockchain network.

Given that transactions are broadcasted in blockchain networks, anyone can receive a piece of desired information in an oblivious way by downloading and screening all transactions. This feature is particularly suitable for secret communication. Meanwhile, to protect confidentiality, data are usually encrypted which makes the data processing difficult on smart contracts [65].

Another promising security research topic is security analysis and enhancement on smart contracts. Such contract is an automatically executed program running on blockchain, and vulnerabilities [294] have been found on them constantly which can cause a massive financial damage.

Privacy   Privacy protection mechanisms stem from the awareness of users' privacy [302–304, 310] concerns. Especially, in finance involved blockchain networks, e.g., Bitcoin and Ethereum, a lot of work have been conducted to protect transaction privacy to some extent. Still, there are not perfectly designed (e.g., identity privacy are still facing linking attacks and locations are not well protected if the service providers collect users' data continuously) and more efforts need to be done before privacy is further protect from de-anonymization attacks. Some new cryptographic techniques, including succinct non-interactive zero-knowledge proof (zk-SNARK), have been used in blockchain to support efficient privacy protection. Blockchain-enabled IoT systems can use this dynamic combination as reference as well.

Blockchain data are likely to be analyzed in other platforms and data owners should be able to request blockchain to remove their data from being put into data analysis, especially when the date is sensitive. But this is not possible to guarantee since blockchain records are immutable and kept forever. Some work mentioned redactable blockchain [295], but this violates the essence of blockchain.

9.3   Summary

This chapter discusses some future research topics in blockchain-enabled IoT systems, including a further integration, managing trust, providing efficient and high-capacity data storage, enabling data analysis, and enhancing data security and user privacy. As a novel and developing technology, blockchain will undoubtedly march towards a mature and user-friendly tool. Moreover, it will continue to demonstrate its power in integrating with IoT systems.

Exercises

9.1 How to further integrate blockchain with IoT applications? Refer to Sect. 9.2.1.

9.2 What causes the trust issue in blockchain-enabled IoT systems? How to address this issue? Refer to Sect. 9.2.2.

9.3 If an efficient and high-capacity data storage system can be built atop blockchain, how will you design it? Refer to Sect. 9.2.3.

9.4 Since a huge amount of transactions are stored on blockchain, is it possible to conduct some data analytics based on the data in transactions? Refer to Sect. 9.2.4.

9.5 How to enhance security and privacy with the help of blockchain? Refer to Sect. 9.2.5.

9.6 Say Alice is designing a new blockchain for her transportation company, what issues other than efficiency and security should she address before the implementation?

Appendix A
Setup for a Local Ethereum Platform

Ethereum [55, 296] is an open-source and blockchain-based distributed platform featuring smart contract functionality. It is a transaction-based state machine and it attempts to provide a firmly integrated end-to-end system to developers. In this section, we introduce how to set up a local Ethereum network.

- We install geth [297], i.e., go-ethereum, which is an Ethereum client end and then install an Ethereum wallet [298].
- We create a file genesis.json in "C:\ethereum" and its source code is shown in Fig. A.1.
- We initiate a genesis block by compiling genesis.json through a command line "geth –datadir "C:\ethereum" init genesis.json" in the geth console as shown in Figs. A.1 and A.2.
- We initiate a private blockchain through a command line "geth -datadir "%cd%\chain" console" in the geth console as shown in Fig. A.3.
- We create a new account with secret key "silian2018" to start mining in the geth console through a command line personal1.newAccount('silian2018') as shown in Fig. A.4.
- To exit, we can use a command line "exit" in the geth console "exit" as shown in Fig. A.5
- To initiate the node in the private blockchain, we can use a command line "geth -targetgaslimit 4294967295 -rpc -rpcaddr "10.108.15.245" -rpcport "8101" -port "30301" -rpcapi "eth,web3,personal" -networkid 2016 -identity 2016 - nodiscover -maxpeers 5 -datadir "%cd%\chain" -unlock 0 -rpccorsdomain "*" -mine console" as shown in Fig. A.6.
- We initiate the Ethereum Wallet console as shown in Fig. A.7.
- We transfer 10 Ether from one account to another account as shown in Fig. A.8.
- We can see the transfer confirmation page in Fig. A.9.
- We can see the balance of each account as shown in Fig. A.10.

© Springer Nature Switzerland AG 2019
L. Zhu et al., *Blockchain Technology in Internet of Things*,
https://doi.org/10.1007/978-3-030-21766-2

```
{
    "nonce":"0x0000000000000042",
    "mixhash":"0x0000000000000000000000000000000000000000000000000000000000000000",
    "difficulty": "0x4000",
    "alloc": {},
    "coinbase":"0x0000000000000000000000000000000000000000",
    "timestamp": "0x00",
    "parentHash":"0x0000000000000000000000000000000000000000000000000000000000000000",
    "extraData": "PICC GenesisBlock",
    "gasLimit":"0xffffffff"
}
```

Fig. A.1 Genesis.json

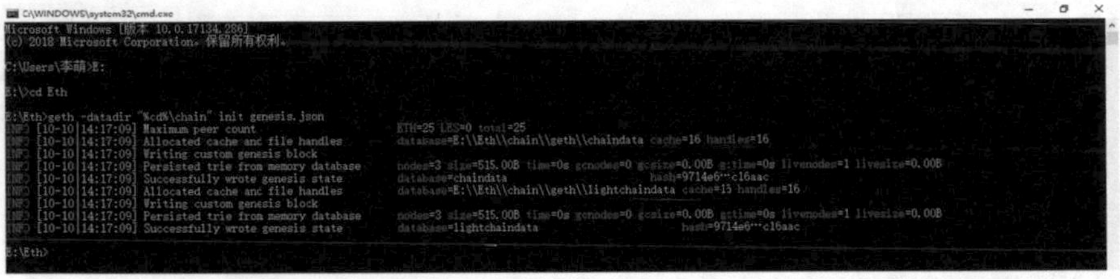

Fig. A.2 Initializing a genesis block

Fig. A.3 Initializing a private blockchain

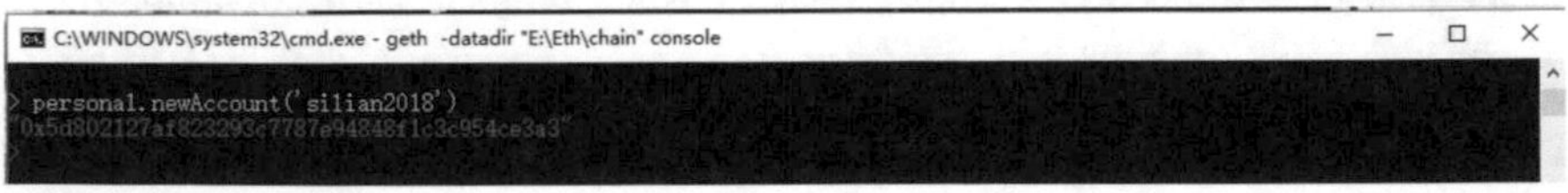

Fig. A.4 Creating a new account

Fig. A.5 Exit

Fig. A.6 Initiating the node in the private blockchain

Fig. A.7 Ethereum wallet initiation

Fig. A.8 Transferring 10 ethers

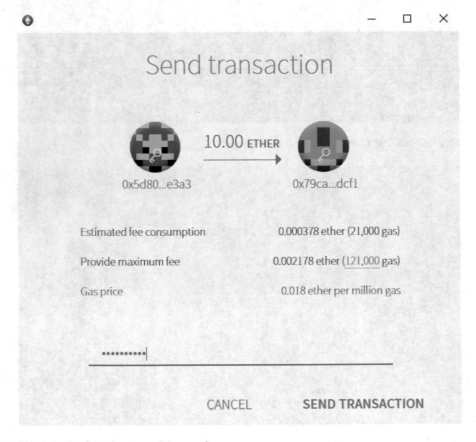

Fig. A.9 Confirmation page of the transfer

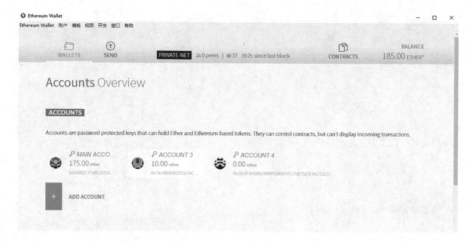

Fig. A.10 Account information

Appendix B
A Sample Examination Paper

1. Which one is the first application of blockchain? (1 point)

   A. Bitcoin
   B. Ethereum
   C. Filecoin
   D. Hyperledger

2. Which one could provide a potential solution for financial systems and industries? (1 point)

   A. Bitcoin
   B. Ethereum
   C. Consortium blockchain
   D. Public blockchain

3. Which layer is the core of blockchain? (1 point)

   A. Application
   B. Consensus
   C. Network
   D. Contract

4. Which one is not a key feature of blockchain? (1 point)

   A. Decentralization
   B. Confidentiality
   C. Tamper resistance
   D. Public

© Springer Nature Switzerland AG 2019   121
L. Zhu et al., *Blockchain Technology in Internet of Things*,
https://doi.org/10.1007/978-3-030-21766-2

5. Which technique does blockchain choose to link the blocks? (1 point)

   A. Encryption
   B. Signature
   C. Hash
   D. Slicing

6. Blockchain is classified into (1 point)

   A. Public blockchain
   B. Private blockchain
   C. Consortium blockchain
   D. Transparent blockchain

7. What can be considered as a consensus algorithm(s) (1 point)?

   A. Proof-of-work
   B. Proof-of-stake
   C. Proof of knowledge
   D. Practical Byzantine Fault Tolerance

8. Which one(s) is the application of IoT? (1 point)

   A. Smart grid
   B. Wireless senor networks
   C. Vehicular networks
   D. Smart home

9. What are security issues in IoT? (1 point)

   A. Confidentiality
   B. Integrity
   C. Authentication
   D. Non-repudiation

10. What can be seen as user privacy in IoT? (1 point)

    A. Identity
    B. Location
    C. Salary
    D. Hospital treatment history

11. Bitcoin has adopted _____ as its signature scheme. (1 point)

12. Block header in Bitcoin includes _____, _____, _____, _____, _____, and _____. (6 points)

13. Bitcoin chooses _____ as its hash function. (1 point)

14. The smallest unit of the Bitcoin currency is called _____. (1 point)

15. The block generation process is called _____ by miners. (1 point)

16. A wallet has two keys which are _____ and _____. (2 points)

17. The generation of a Bitcoin block and an Ethereum block usually takes _____ and _____, respectively. (2 points)

18. 18. _____ will be used in executing a transaction in Ethereum. (1 point)

19. Name four items in an Ethereum block: _____, _____, _____, and _____. (4 points)

20. In a transaction, the total value of inputs should _____ the total value of outputs.

21. What is the difference between soft fork and hard fork? (5 points)

22. What are the strategies of relaying transactions in Bitcoin blockchain? (5 points)

23. What is an uncle block? What are the benefits of introducing uncle blocks? (5 points)

24. What are the basic components in vehicular networks? And what are pertinent security and privacy issues? (5 points)

25. How to store data with IoT blockchain and what are the side effects? (5 points)

26. How to choose consensus algorithms for different IoT applications? (5 points)

27. Please give two examples for each classification of blockchain and explain how does blockchain fit in these scenarios. (10 points)

28. What are the principles of combining blockchain with IoT applications? (10 points)

29. What are the common techniques to protect security in IoT? (10 points)

30. What is privacy? And please give three examples in IoT and how to protect them. (10 points)

Appendix C
Project 1: Blockchain for Supply Chain

Now consider an application that needs to monitor details of all the food in a supply chain with a shared, trustworthy, secure record. Please design a blockchain-enabled supply chain system. Specific requirements are as follows:

- Including entities: food sources, transporters, factories, retailers, and customers.
- What needs to be recorded: the status of food, the location and flow of food, sender and receiver of food flow, the environmental conditions of food preservation and transportation.
- Designing a blockchain: construct a proper blockchain for the supply chain system.
- Protecting security and privacy: name a few security and privacy aspects that need to be protected, explain why, and enforce corresponding protection methods.
- Guaranteeing efficiency: achieve light-weight operations and blockchain maintenance for all entities in supply chain.

© Springer Nature Switzerland AG 2019
L. Zhu et al., *Blockchain Technology in Internet of Things*,
https://doi.org/10.1007/978-3-030-21766-2

Appendix D
Project 2: Blockchain for Collaborative Trade Management

Now consider a group of collaborative companies that need to share and record trade information (e.g., tea, liquor) among them in an internal ledger. Please design a blockchain-enabled data sharing system. Specific requirements are as follows:

- Including entities: three companies.
- What needs to be recorded: the type of data, timestamp of each transaction, sender and receiver of each transaction.
- Designing a blockchain: construct a proper blockchain for the data sharing system.
- Protecting security and privacy: name a few security and privacy aspects that need to be protected, explain why, and enforce corresponding protection methods.
- Guaranteeing efficiency: achieve light-weight operations and blockchain maintenance for the three companies.

© Springer Nature Switzerland AG 2019
L. Zhu et al., *Blockchain Technology in Internet of Things*,
https://doi.org/10.1007/978-3-030-21766-2

References

1. S. Nakamoto, Bitcoin: a peer-to-peer electronic cash system (2009). Available: https://bitcoin. org/bitcoin.pdf
2. K. Gai, M. Qiu, X. Sun, A survey on FinTech. J. Netw. Comput. Appl. **103**, 262–273 (2018)
3. K. Gai, M. Qiu, B. Thuraisingham, L. Tao, Proactive attribute-based secure data schema for mobile cloud in financial industry, in *IEEE 17th International Conference on High Performance Computing and Communications (HPCC)* (IEEE, Piscataway, 2015), pp. 1332–1337
4. K. Gai, M. Qiu, X. Sun, H. Zhao, Security and privacy issues: a survey on FinTech, in *International Conference on Smart Computing and Communication* (Springer, Cham, 2016), pp. 236–247
5. Preparing for a Blockchain Future (2018). Available: https://sloanreview.mit.edu/article/ preparing-for-a-blockchain-future
6. L. Atzori, A. Iera, G. Morabito, The Internet of Things: a survey. Comput. Netw. **54**(15), 2787–2805 (2010)
7. K. Gai, M. Qiu, Z. Xiong, M. Liu, Privacy-preserving multi-channel communication in edge-of-things. Futur. Gener. Comput. Syst. **85**, 190–200 (2018)
8. K. Gai, K.-K.R. Choo, M. Qiu, L. Zhu, Privacy-preserving content-oriented wireless communication in Internet-of-Things. IEEE Internet Things J. **5**(4), 3059–3067 (2018)
9. X. He, C. Wang, T. Liu, K. Gai, D. Chen, L. Bai, Research on campus mobile model based on periodic purpose for opportunistic network, in *IEEE 17th International Conference on High Performance Computing and Communications (HPCC)* (IEEE, Piscataway, 2015), pp. 782–785
10. L. Tao, S. Golikov, K. Gai, M. Qiu, A reusable software component for integrated syntax and semantic validation for services computing, in *2015 IEEE Symposium on Service-Oriented System Engineering (SOSE)* (IEEE, Piscataway, 2015), pp. 127–132
11. M. Li, L. Zhu, Z. Zhang, X. Du, M. Guizani, PROS: a privacy-preserving route sharing service via vehicular fog computing. IEEE Access **6**, 66188–66197 (2018)
12. H. Li, L. Zhu, M. Shen, F. Gao, X. Tao, S. Liu, Blockchain-based data preservation system for medical data. J. Med. Syst. **42**(141), 1–13 (2018)
13. K. Gai, M. Qiu, H. Zhao, Security-aware efficient mass distributed storage approach for cloud systems in big data, in *IEEE 2nd International Conference on Big Data Security on Cloud (BigDataSecurity)* (IEEE, Piscataway, 2016), pp. 140–145
14. K. Gai, A. Steenkamp, A feasibility study of platform-as-a-service using cloud computing for a global service organization. J. Inf. Syst. Appl. Res. **7**(3), 28 (2014)

© Springer Nature Switzerland AG 2019

L. Zhu et al., *Blockchain Technology in Internet of Things*,

https://doi.org/10.1007/978-3-030-21766-2

15. K. Gai, M. Qiu, L.-C. Chen, M. Liu, Electronic health record error prevention approach using ontology in big data, in *IEEE 17th International Conference on High Performance Computing and Communications (HPCC)* (IEEE, Piscataway, 2015), pp. 752–757

16. K. Gai, M. Qiu, S. Jayaraman, L. Tao, Ontology-based knowledge representation for secure self-diagnosis in patient-centered teleheath with cloud systems, in *2015 IEEE 2nd International Conference on Cyber Security and Cloud Computing (CSCloud)* (IEEE, Piscataway, 2015), pp. 98–103

17. K. Thakur, M.L. Ali, K. Gai, M. Qiu, Information security policy for e-commerce in Saudi Arabia, in *IEEE 2nd International Conference on Big Data Security on Cloud (BigDataSecurity)* (IEEE, Piscataway, 2016), pp. 187–190

18. M. Sette, L. Tao, K. Gai, N. Jiang, A semantic approach to intelligent and personal tutoring system, in *2016 IEEE 3rd International Conference on Cyber Security and Cloud Computing (CSCloud)* (IEEE, Piscataway, 2016), pp. 261–266

19. L. Qiu, K. Gai, M. Qiu, Optimal big data sharing approach for tele-health in cloud computing, in *IEEE International Conference on Smart Cloud (SmartCloud)* (IEEE, Piscataway, 2016), pp. 184–189

20. J. Bonneau, A. Miller, J. Clark, A. Narayanan, J.A. Kroll, E.W. Felten, SoK: research perspectives and challenges for bitcoin and cryptocurrencies, in *Proceedings of the IEEE Symposium on Security and Privacy (S&P)* (2015), pp. 104–121

21. A. Narayanan, J. Bonneau, E. Felten, A. Miller, S. Goldfeder, Bitcoin and cryptocurrency technologies. Available: http://www.freetechbooks.com/bitcoin-and-cryptocurrency-technologies-t938.html (2016), pp. 1–308

22. Z. Zhang, S. Xie, H. Dai, X. Chen, H. Wang, Blockchain challenges and opportunities: a survey. Int. J. Web Grid Serv. **14**(4), 352 (2017)

23. M.E. Peck, Blockchains how they work and why they'll change the world (2017). Available: https://spectrum.ieee.org/computing/networks/blockchains-how-they-work-and-why-theyll-change-the-world

24. T.T.A. Dinh, R. Liu, M. Zhang, G. Chen, B.C. Ooi, J. Wang, Untangling blockchain a data processing view of blockchain systems. IEEE Trans. Knowl. Data Eng. **30**(7), 1366–1385 (2018)

25. M. Li, L. Zhu, X. Lin, Efficient and privacy-preserving carpooling using blockchain-assisted vehicular fog computing. IEEE Internet Things J. **PP**(99), 1–13 (2018)

26. Z. Yang, K. Yang, L. Lei, K. Zheng, V.M. Leung, Blockchain-based decentralized trust management in vehicular networks. IEEE Internet Things J. **6**(2), 1495–1505 (2019)

27. D. Tse, B. Zhang, Y. Yang, C. Cheng, H. Mu, Blockchain application in food supply information security, in *Proceedings of the IEEE International Conference on Industrial Engineering and Engineering Management (IEEM)* (2017), pp. 1357–1360

28. Q. Xia, E.B. Sifah, K.O. Asamoah, J. Gao, X. Du, M. Guizani, MeDShare: trust-less medical data sharing among cloud service providers via blockchain. IEEE Access **5**, 14757–14767 (2017)

29. E. Androulaki et al., Hyperledger fabric: a distributed operating system for permissioned blockchains, in *Proceedings of the 13th European Conference on Computer Systems (EuroSys)* (2018), pp. 1–15

30. Corda (2019). Available: http://www.corda.net

31. Z. Li, J. Kang, R. Yu, D. Ye, Q. Deng, Y. Zhang, Consortium blockchain for secure energy trading in industrial Internet of Things. IEEE Trans. Ind. Inf. (TII) **14**(8), 1690–1700 (2018)

32. C. Garman, M. Green, I. Miers, Accountable privacy for decentralized anonymous payments, in *International Conference on Financial Cryptography & Data Security (FC)* (2016), pp. 1–28

33. A. Kosba, A. Miller, E. Shi, Z. Wen, C. Papamanthou, Hawk: the blockchain model of cryptography and privacy-preserving smart contracts, in *Proceedings of the IEEE 37th Symposium on Security and Privacy (S&P)* (2016), pp. 839–858

34. A. Dorri, M. Steger, S.S. Kanhere, R. Jurdak, BlockChain: a distributed solution to automotive security and privacy. IEEE Commun. Mag. **15**(12), 119–125 (2017)

35. S. Hu, C. Cai, Q. Wang, C. Wang, X. Luo, K. Ren, Searching an encrypted cloud meets blockchain: a decentralized, reliable and fair realization, in *Proceedings of the 37th IEEE Conference on Computer Communications (INFOCOM)* (2018), pp. 1–9

36. K. Gai, M. Qiu, H. Zhao, J. Xiong, Privacy-aware adaptive data encryption strategy of big data in cloud computing, in *2016 IEEE 3rd International Conference on Cyber Security and Cloud Computing (CSCloud)* (IEEE, Piscataway, 2016), pp. 273–278

37. M. Andrychowicz, S. Dziembowski, D. Malinowski, Ł. Mazurek, Secure multiparty computations on bitcoin, in *2014 IEEE Symposium on Security and Privacy* (2014), pp. 443–458

38. A. Ouaddah, A.A. Elkalam, A.A. Ouahman, FairAccess: a new blockchain-based access control framework for the Internet of Things. Secur. Commun. Netw. **9**(18), 5943–5964 (2017)

39. L. Ma, L. Tao, Y. Zhong, K. Gai, RuleSN: research and application of social network access control model, in *IEEE 2nd International Conference on Big Data Security on Cloud (BigDataSecurity)* (IEEE, Piscataway, 2016), pp. 418–423

40. L. Ma, L. Tao, K. Gai, Y. Zhong, A novel social network access control model using logical authorization language in cloud computing. Concurrency Comput Pract. Experience **29**(14), e3893 (2017)

41. S. Azouvi, M. Al-Bassam, S. Meiklejohn, Who am I? Secure identity registration on distributed ledgers, in *International Workshop on Data Privacy Management* (2017), pp. 373–389

42. L. Atzori, A. Iera, G. Morabito, The Internet of Things: a survey. Comput. Netw. **54**(15), 2787–2805 (2010)

43. J. Gubbi, R. Buyya, S. Marusic, M. Palaniswami, Internet of Things (IoT): a vision, architectural elements, and future directions. Futur. Gener. Comput. Syst. **29**(7), 1645–1660 (2013)

44. E. Borgia, The Internet of Things vision: key features, applications and open issues. Comput. Commun. **54**, 1–31 (2014)

45. A. Al-Fuqaha, M. Guizani, M. Mohammadi, M. Aledhari, M. Ayyash, Internet of Things: a survey on enabling technologies, protocols, and applications. IEEE Commun. Surv. Tutorials **17**(4), 2347–2376 (2015)

46. K. Gai, M. Qiu, H. Zhao, L. Tao, Z. Zong, Dynamic energy-aware cloudlet-based mobile cloud computing model for green computing. J. Netw. Comput. Appl. **59**, 46–54 (2016)

47. K. Gai, S. Li, Towards cloud computing: a literature review on cloud computing and its development trends, in *2012 Fourth International Conference on Multimedia Information Networking and Security (MINES)* (IEEE, Piscataway, 2012), pp. 142–146

48. K. Gai, M. Qiu, Z. Ming, H. Zhao, L. Qiu, Spoofing-jamming attack strategy using optimal power distributions in wireless smart grid networks. IEEE Trans. Smart Grid **8**(5), 2431–2439 (2017)

49. K. Gai, J. Pan, Human resource management: a case study of the air traffic controller strike in 1981. China Manag. Informatization **12**(15), 61–65 (2009)

50. K. Gai, M. Qiu, H. Zhao, W. Dai, Privacy-preserving adaptive multi-channel communications under timing constraints, in *IEEE International Conference on Smart Cloud (SmartCloud)* (IEEE, Piscataway, 2016), pp. 190–195

51. K. Gai, M. Qiu, H. Zhao, L. Qiu, Smart energy-aware data allocation for heterogeneous memory, in *IEEE 18th International Conference on High Performance Computing and Communications (HPCC)* (IEEE, Piscataway, 2016), pp. 136–143

52. K. Gai, A report about suggestions on developing e-learning in China, in *2010 International Conference on E-Business and E-Government (ICEE)* (IEEE, Piscataway, 2010), pp. 609–613

53. K. Gai, M. Qiu, H. Zhao, X. Sun, Resource management in sustainable cyber-physical systems using heterogeneous cloud computing. IEEE Trans. Sustain. Comput. **3**(2), 60–72 (2018)

54. Transactions (2019). Available: https://bitcoin.org/en/developer-guide#transactions

55. Ethereum (2019). Available: https://www.ethereum.org

56. Ethereum: a secure decentralised generalised transaction ledger byzantium version 69351d5 (2018). Available: https://ethereum.github.io/yellowpaper/paper.pdf
57. Block Height and Forking (2019). Available: https://bitcoin.org/en/developer-guide#block-height-and-forking
58. Multichain (2019). Available: https://www.multichain.com/
59. eosio - The most powerful infrastructure for decentralized applications (2019). Available: https://eos.io
60. BTM Blockchain (2019). Available: https://www.blockchaintmhub.io
61. GXChain - rebuild credit society with blockchain designed to build a trusted data internet of value (2019). Available: https://www.gxb.io/en
62. A free and lightning fast peer-to-peer transactional network for digital assets (2019). Available: https://mixin.one
63. Adaptable blockchain for enterprise solutions (2019). Available: https://nuls.io
64. Commercial blockchain platform anti-counterfeiting traceability application pioneer (2019). Available: https://www.inchain.org
65. N. Szabo, Formalizing and securing relationships on public networks. First Monday **2**(9) (1997)
66. K. Christidis, M. Devetsikiotis, Blockchains and smart contracts for the Internet of Things. IEEE Access **4**, 2292–2303 (2016)
67. Empower my supply chain. Skuchain. Available: http://www.skuchain.com
68. Koinify Raises $1 Million for Smart Corporation Crowdfunding Platform. Available: https://www.coindesk.com/koinify-1-million-smart-corporation-crowdfunding
69. A.K.R. Dermody, O. Slama, Counterparty announcement. Available: https://bitcointalk.org/index.php?topic=395761.0
70. SpankChain loses $40K in hack due to smart contract bug. Available: https://www.coindesk.com/spankchain-loses-40k-in-hack-due-to-smart-contract-bug
71. Researchers find 34,200 vulnerable ethereum smart contracts. Available: https://www.bleepingcomputer.com/news/cryptocurrency/researchers-find-34-200-vulnerable-ethereum-smart-contracts
72. L. Lamport, R. Shostak, M. Pease, The byzantine generals problem. ACM Trans. Program. Lang. Syst. **4**(3), 382–401 (1982)
73. Primecoin: a new type cryptocurrency which is Proof-of-Work based on searching for prime numbers (2019). http://primecoin.io
74. I. Bentov, C. Lee, A. Mizrahi, M. Rosenfeld, Proof of activity: extending bitcoin's proof of work via proof of stake [extended abstract]. ACM SIGMETRICS Perform. Eval. Rev. **42**(3), 34–37 (2014)
75. A. Kiayias, A. Russell, B. David, R. Oliynykov, Ouroboros: a provably secure proof-of-stake blockchain protocol, in *International Cryptology Conference* (2017), pp. 357–388
76. V. Zamfir, Introducing Casper "the Friendly Ghost" (2015). Available: https://blog.ethereum.org/2015/08/01/introducing-casper-friendly-ghost
77. M. Castro, B. Liskov, Practical Byzantine fault tolerance, in *Proceedings of the 3rd Symposium on Operating Systems Design and Implementation* (1999), pp. 1–14
78. B. Curran, What is Practical Byzantine Fault Tolerance? Complete Beginner's Guide (2018). Available: https://blockonomi.com/practical-byzantine-fault-tolerance
79. BITSHARE BLOCKCHAIN (2019). Available: https://bitshares.org
80. F. Tian, An agri-food supply chain traceability system for China based on RFID & blockchain technology, in *Proceedings of the 13th International Conference on Service Systems and Service Management (ICSSSM)* (2016), pp. 1–6
81. Fresh duck eggs found to contain Sudan Red in Fujian (2006). Available: http://www.china.org.cn/english/health/189567.htm
82. Horsemeat scandal: the essential guide (2013). Available: https://www.theguardian.com/uk/2013/feb/15/horsemeat-scandal-the-essential-guide

83. A supply chain traceability system for food safety based on HACCP, blockchain & Internet of Things, in *Proceedings of the 14th International Conference on Service Systems and Service Management (ICSSSM)* (2017), pp. 1–6

84. I. Allison, Meet BigchainDB - 'the scalable blockchain database' hitting one million writes per second (2016). Available: https://www.ibtimes.co.uk/meet-bigchaindb-scalable-blockchain-database-hitting-one-million-writes-per-second-1544918

85. H. Yin, K. Gai, An empirical study on preprocessing high-dimensional class-imbalanced data for classification, in *IEEE 17th International Conference on High Performance Computing and Communications (HPCC)* (IEEE, Piscataway, 2015), pp. 1314–1319

86. N. Koshizuka, K. Sakamura, Ubiquitous ID: standards for ubiquitous computing and the Internet of Things. IEEE Pervasive Comput. **9**(4), 98–101 (2010)

87. Belkin (2019). Available: https://www.belkin.com/us/Products/smarthome-iot/c/wemo

88. P. Levis, S. Madden, J. Polastre, R. Szewczyk, K. Whitehouse, A. Woo, D. Gay, J. Hill, M. Welsh, E. Brewer, D. Culler, TinyOS: an operating system for sensor networks, in *Ambient Intelligence* (Springer, Berlin, 2005), pp. 115–148

89. E. Baccelli, O. Hahm, M. Günes, M. Wählisch, T.C. Schmidt, RIOT OS: towards an OS for the Internet of Things, in *Proceedings of the IEEE Conference on Computer Communications (INFOCOM) Workshops* (2013), pp. 79–80

90. Y. Li, K. Gai, M. Qiu, W. Dai, M. Liu, Adaptive human detection approach using FPGA-based parallel architecture in reconfigurable hardware. Concurrency Comput. Pract. Experience **29**(14), e3923 (2017)

91. H. Chan, A. Perrig, D. Song, Secure hierarchical in-network aggregation for sensor networks, in *ACM Conference on Computer and Communications Security (CCS)* (2006), pp. 1–10

92. L. Zhu, M. Li, DoS-resilient secure data aggregation in wireless sensor networks, in *Poster of the17th ACM Annual International Conference on Mobile Computing and Networking (ACM MobiCom)* (2011), pp. 1–2

93. An energy efficient and integrity-preserving aggregation protocol in wireless sensor networks, in *Proceedings of the IEEE International Performance Computing and Communications Conference (IPCCC)*, December 2011, pp. 1–6

94. L. Zhu, Z. Yang, M. Li, D. Liu, An efficient data aggregation protocol concentrated on data integrity in wireless sensor networks. Int. J. Distrib. Sens. Netw. **2013**(7), 1–9 (2013)

95. L. Zhu, Z. Yang, M. Wang, M. Li, ID list forwarding free confidentiality preserving data aggregation for wireless sensor networks. Int. J. Distrib. Sens. Netw. **2013**(241261), 59–72 (2014)

96. K. Gai, M. Qiu, H. Zhao, M. Liu, Energy-aware optimal task assignment for mobile heterogeneous embedded systems in cloud computing, in *2016 IEEE 3rd International Conference on Cyber Security and Cloud Computing (CSCloud)* (IEEE, Piscataway, 2016), pp. 198–203

97. M. Qiu, D. Cao, H. Su, K. Gai, Data transfer minimization for financial derivative pricing using Monte Carlo simulation with GPU in 5G, Int. J. Commun. Syst. **29**(16), 2364–2374 (2016)

98. X. Xing, J. Wang, M. Li, Services and key technologies of the Internet of Things. ZTE Commun. **2**, 1–11 (2010)

99. M. Gigli, S. Koo, Internet of Things: services and applications categorization. Adv. Internet Things **1**(2), 27–31 (2011)

100. R. Khan, S.U. Khan, R. Zaheer, S. Khan, Future Internet: the Internet of Things architecture, possible applications and key challenges, in *Proceedings of the 10th International Conference on Frontiers of Information Technology* (2012), pp. 257–260

101. M. Qiu, Z. Ming, J. Li, K. Gai, Z. Zong, Phase-change memory optimization for green cloud genetic algorithm. IEEE Trans. Comput. **64**(12), 3528–3540 (2015)

102. H. Zhao, M. Chen, M. Qiu, K. Gai, M. Liu, A novel pre-cache schema for high performance android system. Futur. Gener. Comput. Syst. **56**, 766–772 (2016)

103. H. Zhao, M. Qiu, K. Gai, J. Li, X. He, Maintainable mobile model using pre-cache technology for high performance android system, in *2015 IEEE 2nd International Conference on Cyber Security and Cloud Computing (CSCloud)* (IEEE, Piscataway, 2015), pp. 175–180

104. K. Gai, M. Qiu, H. Zhao, Cost-aware multimedia data allocation for heterogeneous memory using genetic algorithm in cloud computing. IEEE Trans. Cloud Comput. (2016)

105. Y. Li, K. Gai, L. Qiu, M. Qiu, H. Zhao, Intelligent cryptography approach for secure distributed big data storage in cloud computing. Inf. Sci. **387**, 103–115 (2017)

106. K. Gai, M. Qiu, H. Zhao, Energy-aware task assignment for mobile cyber-enabled applications in heterogeneous cloud computing. J. Parallel Distrib. Comput. **111**, 126–135 (2018)

107. K. Gai, L. Qiu, M. Chen, H. Zhao, M. Qiu, Sa-east: security-aware efficient data transmission for its in mobile heterogeneous cloud computing. ACM Trans. Embed. Comput. Syst. **16**(2), 60 (2017)

108. K. Gai, A. Steenkamp, Feasibility of a platform-as-a-service implementation using cloud computing for a global service organization, in *Proceedings of the Conference for Information Systems Applied Research*, vol. 2167 (2013), p. 1508

109. L. Chen, Y. Duan, M. Qiu, J. Xiong, K. Gai, Adaptive resource allocation optimization in heterogeneous mobile cloud systems, in *2015 IEEE 2nd International Conference on Cyber Security and Cloud Computing (CSCloud)* (IEEE, Piscataway, 2015), pp. 19–24

110. K. Gai, A review of leveraging private cloud computing in financial service institutions: value propositions and current performances. Int. J. Comput. Appl. **95**(3), 40–44 (2014)

111. Propel research with big data and breakthrough insights (2019). Available: https://edu.google.com/products/google-cloud-platform/?modal_active=none

112. Data Archive (2019). Available: https://aws.amazon.com/archive/?nc2=h_m2

113. F. Bonomi, R.A. Milito, J. Zhu, S. Addepalli, Fog computing and its role in the Internet of Things, in *Proceedings of the Workshop on Mobile Cloud Computing (MCC)* (2012), pp. 13–16

114. J. Shropshire, Extending the cloud with fog: security challenges and opportunities, in *Proceedings of Information System* (2014), pp. 1–10

115. J. Ni, K. Zhang, X. Lin, X. Shen, Securing fog computing for Internet of Things applications: challenges and solutions. IEEE Commun. Surv. Tutorials **20**(1), 601–628 (2017)

116. C. Sheng, Y. Chang, J. Lin, M. Li, Z. Zhang, L. Zhu, A residential privacy mining scheme based on smart meter readings, in *Proceedings of the 3rd International Symposium on Privacy Computing (PriCom)*, November 2017, pp. 1–6

117. C. Asamoah, L. Tao, K. Gai, N. Jiang, Powering filtration process of cyber security ecosystem using knowledge graph, in *2016 IEEE 3rd International Conference on Cyber Security and Cloud Computing (CSCloud)* (IEEE, Piscataway, 2016), pp. 240–246

118. Z. Zhang, Z. Qin, L. Zhu, J. Weng, K. Ren, Cost-friendly differential privacy for smart meters: exploiting the dual roles of the noise. IEEE Trans. Smart Grid **8**(2), 619–626 (2017)

119. L. Zhu, M. Li, Z. Zhang, Q. Zhan, ASAP: an anonymous smart-parking and payment scheme in vehicular networks. IEEE Trans. Dependable Secure Comput. **PP**(99), 1–12 (2018)

120. A. Butowsky, K. Gai, M. Coakley, M. Qiu, C.C. Tappert, City of white plains parking app: case study of a smart city web application, in *2015 IEEE 2nd International Conference on Cyber Security and Cloud Computing (CSCloud)* (IEEE, Piscataway, 2015), pp. 278–282

121. L. Zhu, Y. Wu, K. Gai, K.K.R. Choo, Controllable and trustworthy blockchain-based cloud data management. Futur. Gener. Comput. Syst. **91**, 527–535 (2019)

122. S. Sicari, A. Rizzardi, L.A. Grieco, A. Coen-Porisini, Security, privacy and trust in Internet of Things: the road ahead. Comput. Netw. **76**, 146–164 (2015)

123. J. Ni, A. Zhang, X. Lin, X. Shen, Security, privacy, and fairness in fog-based vehicular crowdsensing. IEEE Commun. Mag. **55**(6), 146–152 (2017)

124. K. Gai, M. Qiu, L. Tao, Y. Zhu, Intrusion detection techniques for mobile cloud computing in heterogeneous 5G. Secur. Commun. Netw. **9**(16), 3049–3058 (2016)

125. K. Gai, M. Qiu, Blend arithmetic operations on tensor-based fully homomorphic encryption over real numbers. IEEE Trans. Ind. Inf. **14**(8), 3590–3598 (2018)

126. K. Gai, M. Qiu, H. Zhao, W. Dai, Anti-counterfeit scheme using Monte Carlo simulation for e-commerce in cloud systems, in *2015 IEEE 2nd International Conference on Cyber Security and Cloud Computing (CSCloud)* (IEEE, Piscataway, 2015), pp. 74–79

127. H. Liang, K. Gai, Internet-based anti-counterfeiting pattern with using big data in China, in *IEEE 17th International Conference on High Performance Computing and Communications (HPCC)* (IEEE, Piscataway, 2015), pp. 1387–1392

128. L. Zhu, M. Li, L. Liao, Dynamic group signature scheme based integrity preserving event report. Sens. Lett. **10**(8), 1785–1791 (2012)

129. P. Su, N. Sun, L. Zhu, Y. Li, R. Bi, M. Li, Z. Zhang, A privacy-preserving and vessel authentication scheme using automatic identification system, in *Proceedings of the 5th International Workshop on Security in Cloud Computing (SCC) in conjunction with the 12th ACM Asia Conference on Computer and Communications Security (ASIACCS)*, April 2017, pp. 1–8

130. L. Zhu, M. Li, Z. Zhang, C. Xu, R. Zhang, X. Du, N. Guizani, Privacy-preserving authentication and data aggregation for fog-based smart grid. IEEE Commun. Mag. **57**(6), 80–85 (2019)

131. S.A. Elnagdy, M. Qiu, K. Gai, Cyber incident classifications using ontology-based knowledge representation for cybersecurity insurance in financial industry, in *2016 IEEE 3rd International Conference on Cyber Security and Cloud Computing (CSCloud)* (IEEE, Piscataway, 2016), pp. 301–306

132. K. Thakur, M. Qiu, K. Gai, M.L. Ali, An investigation on cyber security threats and security models, in *2015 IEEE 2nd International Conference on Cyber Security and Cloud Computing (CSCloud)* (IEEE, Piscataway, 2015), pp. 307–311

133. S.A. Elnagdy, M. Qiu, K. Gai, Understanding taxonomy of cyber risks for cybersecurity insurance of financial industry in cloud computing, in *2016 IEEE 3rd International Conference on Cyber Security and Cloud Computing (CSCloud)* (IEEE, Piscataway, 2016), pp. 295–300

134. K. Gai, Y. Wu, L. Zhu, L. Xu, Y. Zhang, Permissioned blockchain and edge computing empowered privacy-preserving smart grid networks. IEEE Internet Things J. **PP**(99), 1–12 (2019)

135. K. Gai, Y. Wu, L. Zhu, M. Qiu, M. Shen, Privacy-preserving energy trading using consortium blockchain in smart grid. IEEE Trans. Ind. Inf. **PP**(99), 1–12 (2019)

136. K. Gai, K.K.R. Choo, L. Zhu, Blockchain-enabled reengineering of cloud datacenters. IEEE Cloud Comput. **5**(6), 21–25 (2018)

137. K. Gai, M. Qiu, Y. Li, X.-Y. Liu, Advanced fully homomorphic encryption scheme over real numbers, in *2017 IEEE 4th International Conference on Cyber Security and Cloud Computing (CSCloud)* (IEEE, Piscataway, 2017), pp. 64–69

138. K. Gai, M. Qiu, An optimal fully homomorphic encryption scheme, in *IEEE International Conference on High Performance and Smart Computing (HPSC), and IEEE International Conference on Intelligent Data and Security (IDS), 2017 IEEE 3rd International Conference on Big Data Security on Cloud (BigDataSecurity)* (IEEE, Piscataway, 2017), pp. 101–106

139. J. Katz, Y. Lindell, *Introduction to Modern Cryptography*, 2nd edn. (CRC Press, Boca Raton, 2015), pp. 1–576

140. L. Zhu, Z. Yang, M. Wang, M. Li, ID list forwarding free confidentiality preserving data aggregation for wireless sensor networks. Int. J. Distrib. Sens. Netw. **2013**, 1–14 (2013)

141. T. Feng, C. Wang, W. Zhang, L. Ruan, Confidentiality protection for distributed sensor data aggregation, in *Proceedings of the 27th IEEE International Conference on Computer Communications (INFOCOM)* (2008), pp. 68–76

142. J. Albath, S. Madria, Secure hierarchical data aggregation in wireless sensor networks, in *Proceedings of the IEEE Wireless Communications and Networking Conference (WCNC)* (2009), pp. 1–6

143. C. Castelluccia, A.C.F. Chan, E. Mykletun, G. Tsudik, Efficient and provably secure aggregation of encrypted data in wireless sensor networks. ACM Trans. Sens. Netw. **5**(3), 1–36 (2009)

144. R. Lu, X. Liang, X. Li, X. Lin, X. Shen, EPPA: an efficient and privacy-preserving aggregation scheme for secure smart grid communications. IEEE Trans. Parallel Distrib. Syst. **23**(9), 1621–1632 (2012)

145. K. Moslehi, R. Kumar, A reliability perspective of the smart grid. IEEE Trans. Smart Grid **1**(1), 57–64 (2010)

146. Z.M. Fadlullah, M.M. Fouda, N. Kato, A. Takeuchi, N. Iwasaki, Y. Nozaki, Toward intelligent machine-to-machine communications in smart grid. IEEE Commun. Mag. **49**(4), 60–65 (2011)

147. D. Boneh, B. Lynn, H. Shacham, Short signatures from the Weil pairing. J. Cryptol. **17**(4), 297–319 (2004)

148. M. Bellare, P. Rogaway, Random oracles are practical: a paradigm for designing efficient protocols, in *Proceedings of the ACM Conference on Computer and Communication Security* (1993), pp. 63–73

149. H. Lee, J. Lee, J. Han, The efficient security architecture for authentication and authorization in the home network, in *3rd International Conference on Natural Computation (ICNC)* (2007), pp. 1–5

150. H. Nicanfar, P. Jokar, V.C.M. Leung, Efficient authentication and key management for the home area network, in *IEEE International Conference on Communications (ICC)* (2012), pp. 8780–882

151. Y.-P. Kim, S. Yoo, C. Yoo, DAoT: dynamic and energy-aware authentication for smart home appliances in Internet of Things, in *IEEE International Conference on Consumer Electronics (ICCE)* (2015), pp. 196–197

152. Voter Privacy: What You Need to Know About Your Digital Trail During the 2016 Election (2016). Available: https://www.eff.org/deeplinks/2016/02/voter-privacy-what-you-need-know-about-your-digital-trail-during-2016-election

153. Privacy Protection in Billing and Health Insurance Communications (2016). Available: https://journalofethics.ama-assn.org/article/privacy-protection-billing-and-health-insurance-communications/2016-03

154. Privacy (2019). Available: https://blinddatewithabook.com/pages/privacy

155. E.D. Cristofaro, C. Soriente, Extended capabilities for a privacy-enhanced participatory sensing infrastructure (PEPSI). IEEE Trans. Inf. Forensics Secur. **8**(12), 2021–2033 (2013)

156. X. Wang, W. Cheng, P. Mohapatra, T. Abdelzaher, Enabling reputation and trust in privacy-preserving mobile sensing. IEEE Trans. Mobile Comput. **13**(12), 2777–2790 (2014)

157. S. Gisdakis, T. Giannetsos, P. Papadimitratos, SPPEAR: security & privacy-preserving architecture for participatory-sensing applications, in *Proceedings of the 7th ACM Conference on Security and Privacy in Wireless and Mobile Networks (WiSec)* (2014), pp. 39–50

158. D. Boneh, X. Boyen, H. Shacham, Short Group Signatures, in *Proceedings of the 24th Annual International Cryptology Conference (CRYPTO)* (2004), pp. 41–55

159. R. Lu, X. Lin, T.H. Luan, X. Liang, X. Shen, Pseudonym changing at social spots: an effective strategy for location privacy in VANETs. IEEE Trans. Veh. Technol. **61**(1), 86–96 (2012)

160. M. Motani, V. Srinivasan, P.S. Nuggehalli, PeopleNet: engineering a wireless virtual social network, in *Proceedings of the 11th Annual International Conference on Mobile Computing and Networking (MobiCom)* (2005), pp. 243–257

161. P. Mohan, V.N. Padmanabhan, R. Ramjee, Nericell: rich monitoring of road and traffic conditions using mobile smartphones, in *Proceedings of the 6th ACM Conference on Embedded Network Sensor Systems (SenSys)* (2008), pp. 357–358

162. Y. Xiao, L. Xiong, Protecting locations with differential privacy under temporal correlations, in *Proceedings of the 22nd ACM SIGSAC Conference on Computer and Communications Security (CCS)* (2015), pp. 1298–1309

163. P. Zhang, C. Hu, D. Chen, H. Li, Q. Li, ShiftRoute: achieving location privacy for map services on smartphones. IEEE Trans. Veh. Technol. **67**(5), 4527–4538 (2018)

164. M. Gruteser, D. Grunwald, Anonymous usage of location-based services through spatial and temporal cloaking, in *Proceedings of the 1st International Conference on Mobile Systems, Applications and Services (MobiSys)* (2003), pp. 31–42

165. H. Kido, Y. Yanagisawa, T. Satoh, An anonymous communication technique using dummies for location-based services, in *Proceedings of the International Conference on Pervasive Services* (2005), pp. 88–97

166. P. Shankar, V. Ganapathy, L. Iftode, Privately querying location-based services with Sybil-Query, in *Proceedings of the 11th International Conference on Ubiquitous Computing (UbiComp)* (2009), pp. 31–40

167. K.C. Lee, W.-C. Lee, H.V. Leong, B. Zheng, Navigational path privacy protection: navigational path privacy protection, in *Proceedings of the 18th ACM Conference on Information and Knowledge Management (CIKM)* (2009), pp. 691–700

168. M.L. Yiu, C.S. Jensen, X. Huang, H. Lu, Spacetwist: managing the trade-offs among location privacy, query performance, and query accuracy in mobile services, in *Proceedings of the IEEE 24th International Conference on Data Engineering (ICDE)* (2008), pp. 366–375

169. S.T. Peddinti, A. Dsouza, N. Saxena, Cover locations: availing location-based services without revealing the location, in *Proceedings of the 10th Annual ACM Workshop on Privacy in the Electronic Society* (2011), pp. 143–152

170. M. Utsunomiya, J. Attanucci, N. Wilson, Potential uses of transit smart card registration and transaction data to improve transit planning. Transp. Res. Rec. J. Transp. Res. Board **1971**(1), 888–896 (2006)

171. Y. Zheng, L.Z. Zhang, X. Xie, W.Y. Ma, Mining interesting locations and travel sequences from GPS trajectories, in *Proceedings of the International Conference on World Wide Web (WWW)* (2009), pp. 791–800

172. X. Cao, G. Cong, C.S. Jensen, Mining significant semantic locations from GPS data, in *Proceedings of the International Conference on Very Large Data Bases (VLDB) Endowment* (2010), pp. 1009–1020

173. B. Agard, C. Morency, M. Trpanierm, Mining public transport user behaviour from smart card data, in *IFAC Symposium on Information Control Problems in Manufacturing (INCOM)*, vol. 12 (2006), pp. 399–404

174. R. Chen, G. Acs, C. Castelluccia, Differentially private sequential data publication via variable-length n-grams, in *ACM Conference on Computer & Communications Security (CCS)* (2012), pp. 638–649

175. C. Dwork, Differential privacy, in *Proceedings of the 33th International Colloquium on Automata, Languages, and Programming (ICALP)*, vol. 4052 (2006), pp. 1–12

176. C. Dwork, F. McSherry, K. Nissim, A. Smith, Calibrating noise to sensitivity in private data analysis, in *Theory of Cryptography Conference (TCC)* (2006), pp. 265–284

177. M. Li, L. Zhu, Z. Zhang, R. Xu, Achieving differential privacy of trajectory data publishing in participatory sensing. Inf. Sci. **400–401**, 1–13 (2017)

178. M. Li, L. Zhu, Z. Zhang, R. Xu, Differentially private publication scheme for trajectory data, in *Proceedings of the 1st IEEE International Conference on Data Science in Cyberspace (DSC)*, June 2016, pp. 596–601

179. M. Li, F. Wu, G. Chen, L. Zhu, Z. Zhang, How to protect query and report privacy without sacrificing service quality in participatory sensing, in *Proceedings of the IEEE International Performance Computing and Communications Conference (IPCCC)*, December 2015, pp. 1–7

180. S. Ding, X. He, J. Wang, B. Qiao, K. Gai, Static node center opportunistic coverage and hexagonal deployment in hybrid crowd sensing. J. Signal Process. Syst. **86**(2–3), 251–267 (2017)

181. Z. Zhang, C. Jin, M. Li, L. Zhu, A perturbed compressed sensing protocol for crowd sensing. Mobile Inf. Syst. (MIS) **2016**, 1–9 (2016)

182. M. Qiu, K. Gai, H. Zhao, M. Liu, Privacy-preserving smart data storage for financial industry in cloud computing. Concurrency Comput. Pract. Experience **30**(5), e4278 (2018)

183. D. Lefeuvre, G. Pavillon, A. Aouba, A. Lamarche-Vadel, A. Fouillet, E. Jougla, G. Rey, Quality comparison of electronic versus paper death certificates in France, 2010. Popul. Health Metr. **12**(3), 18 (2014)

184. I. Oskolkov, R. Shishkov, Converting paper invoice to electronic form for processing of electronic payment thereof, 2014, US Patent 8635156

185. F. Berman, Got data?: a guide to data preservation in the information age. Commun. ACM **51**(12), 50–56 (2008)
186. A. Miller, A. Juels, E. Shi, B. Parno, J. Katz, Permacoin: repurposing bitcoin work for data preservation, in *2014 IEEE Symposium on Security and Privacy* (2014), pp. 475–490
187. M. Swan, *Blockchain: Blueprint for a New Economy* (O'Reilly, Sebastopol, 2015)
188. D.A. Wijaya, Extending asset management system functionality in bitcoin platform, in *2016 International Conference on Computer, Control, Informatics and its Applications (IC3INA)* (2016), pp. 97–101
189. S. Nakamoto, Bitcoin: a peer-to-peer electronic cash system (2008). Available: https://www.bitcoin.org/en/bitcoin-paper
190. S. Bengtsson, B. Solheim, Enforcement of data protection, privacy and security in medical informatics. MEDINFO **92**, 6–10 (1992)
191. J. He, Z. Zhang, M. Li, L. Zhu J. Hu, Provable data integrity of cloud storage service with enhanced security in Internet of Things. IEEE Access **7**, 6226–6239 (2018)
192. Barnaby Jack Could Hack Your Pacemaker and Make Your Heart Explode (2013). Available: https://www.vice.com/en_ca/article/avnx5j/i-worked-out-how-to-remotely-weaponise-a-pacemaker
193. L.M. Arent, R.D. Brownstone, W.A. Fenwick, Ediscovery: preserving, requesting & producing electronic information. Santa Clara Comput. High Technol. Law J. **19**, 131–140 (2002)
194. Example transaction cost. Available: http://ethdocs.org/en/latest/contracts-and-transactions/account-types-gas-and-transactions.html#example-transaction-cost
195. ethereum/go-ethereum: official go implementation of the ethereum protocol (2019). Available: https://github.com/ethereum/go-ethereum
196. Bitcoin, Ethereum, and Litecoin price charts - Coinbase (2019). Available: https://www.coinbase.com/charts
197. Myetherwallet.com (2019). Available: https://www.myetherwallet.com/helpers.html
198. S. Cha, J. Chen, C. Su, K. Yeh, A blockchain connected gateway for BLE-based devices in the Internet of Things. IEEE Access **6**, 24639–24649 (2018)
199. Z. Zhang, W. Cao, Z. Qin, L. Zhu, Z. Yu, K. Ren, When privacy meets economics: enabling differentially-private battery-supported meter reporting in smart grid, in *Proceedings of the IEEE/ACM 25th International Symposium on Quality of Service (IWQoS)* (2017), pp. 1–9
200. G. Liang, S. Weller, F. Luo, J. Zhao, Z. Dong, Distributed blockchain-based data protection framework for modern power systems against cyber attacks. IEEE Trans. Smart Grid **12**(3), 3162–3173 (2018)
201. X. Liu, W. Wang, D. Niyato, N. Zhao, P. Wang, Evolutionary game for mining pool selection in blockchain networks. IEEE Wirel. Commun. Lett. **7**(5), 760–763 (2018)
202. R. Chen, A traceability chain algorithm for artificial neural networks using T-S fuzzy cognitive maps in blockchain. Futur. Gener. Comput. Syst. **80**, 198–210 (2018)
203. X. Yue, H. Wang, D. Jin, M. Li, W. Jiang, Healthcare data gateways: found healthcare intelligence on blockchain with novel privacy risk control. J. Med. Syst. **40**(218), 1–8 (2016)
204. C. Esposito, A. De Santis, G. Tortora, H. Chang, K.K.R. Choo, Blockchain: a panacea for healthcare cloud-based data security and privacy? IEEE Cloud Comput. **5**(1), 31–37 (2018)
205. H. Zhao, M. Qiu, K. Gai, Empirical study of data allocation in heterogeneous memory, in *International Conference on Smart Computing and Communication* (Springer, Berlin, 2017), pp. 385–395
206. G. Zyskind, O. Nathan, Decentralizing privacy: using blockchain to protect personal data, in *Proceedings of the IEEE Security and Privacy Workshops* (2015), pp. 180–184
207. E. Heilman, F. Baldimtsi, S. Goldberg, Blindly signed contracts: anonymous on-blockchain and off-blockchain bitcoin transactions, in *Proceedings of the International Conference on Financial Cryptography and Data Security (FC)* (2016), pp. 43–60
208. X. Li, P. Jiang, T. Chen, X. Luo, Q. Wen, A survey on the security of blockchain systems. Futur. Gener. Comput. Syst. **PP**, 1–13 (2017)
209. L. Zhu, Y. Wu, K. Gai, K.-K.R. Choo, Controllable and trustworthy blockchain-based cloud data management. Futur. Gener. Comput. Syst. **91**, 527–535 (2019)

210. E. Sortomme et al., Optimal scheduling of vehicle-to-grid energy and ancillary services. IEEE Trans. Smart Grid **3**(1), 351–359 (2012)
211. Z.M. Fadlullah et al., GTES: an optimized game-theoretic demand-side management scheme for smart grid. IEEE Syst. J. **8**(2), 588–597 (2013)
212. Y. Wu et al., Optimal pricing and energy scheduling for hybrid energy trading market in future smart grid. IEEE Trans. Ind. Inf. **11**(6), 1585–1596 (2015)
213. Z. Yang et al., P^2: privacy-preserving communication and precise reward architecture for v2g networks in smart grid. IEEE Trans. Smart Grid **2**(4), 697–706 (2011)
214. H. Wang et al., TPP: traceable privacy-preserving communication and precise reward for vehicle-to-grid networks in smart grids. IEEE Trans. Inf. Forensics Secur. **10**(11), 2340–2351 (2015)
215. H.A. Man et al., A new payment system for enhancing location privacy of electric vehicles. IEEE Trans. Veh. Technol. **63**(1), 3–18 (2013)
216. S. Nakamoto, Bitcoin: a peer-to-peer electronic cash system (2009). Available: https://bitcoin.org/bitcoin.pdf
217. Hyperledger/fabric-chaintool. https://github.com/hyperledger/fabric-chaintool
218. D. Zhang, T. He, Y. Liu, S. Lin, J.A. Stankvic, A carpooling recommendation system for taxicab services. IEEE Trans. Emerg. Top. Comput. **2**(3), 254–266 (2014)
219. I.B.-A. Hartman, D. Keren, A.A. Dbai, E. Cohen, L. Knapen, A.-U.-H. Yasar, D. Janssens, Theory and practice in large carpooling problems. Procedia Comput. Sci. **32**(1), 339–347 (2014)
220. B. Caulfield, Estimating the environmental benefits of ride-sharing: a case study of Dublin. Transp. Res. Part D **14**(7), 527–531 (2009)
221. uberPOOL (2019). Available: https://www.uber.com/en-SG/drive/singapore/resources/uberpool
222. Didi Chuxing (2019). Available: https://www.didiglobal.com
223. Y. Zheng, M. Li, W. Lou, Y.T. Hou, Location based handshake and private proximity test with location tags. IEEE Trans. Dependable Secure Comput. **14**(4), 406–419 (2017)
224. R. Li, A.X. Liu, A.L. Wang, B. Bruhadeshwar, Fast range query processing with strong privacy protection for cloud computing, in *Proceedings of the 40th International Conference on Very Large Data Base* (2014), pp. 1953–1964
225. Hyperledger Whitepaper (2017). Available: https://www.yumpu.com/xx/document/view/55615753\hyperledger-whitepaper
226. The difference between public and private blockchain (2017). Available: https://www.ibm.com/blogs/blockchain/2017/05/the-difference-between-public-and-private-blockchain
227. Y. Zhang, C. Xu, S. Yu, H. Li, X. Zhang, SCLPV: secure certificateless public verification for cloud-based cyber-physical-social systems against malicious auditors. IEEE Trans. Comput. Soc. Syst. **2**(4), 159–170 (2015)
228. A.B.T. Sherif, K. Rabieh, M.M.E.A. Mahmoud, X. Liang, Privacy-preserving ride sharing scheme for autonomous vehicles in big data era. IEEE Internet Things J. **4**(2), 611–618 (2016)
229. W. He, X. Liu, M. Ren, Location cheating: a security challenge to location-based social network services, in *Proceedings of the 31st International Conference on Distributed Computing Systems* (2011), pp. 740–749
230. Drivers of Ride-Sharing Services Died after Being Attacked by Passengers (2017). Available: http://lawyersfavorite.com/criminal-attorney/drivers-ride-sharing-services-died-attacked-passengers
231. A. Pham, I. Dacosta, G. Endignoux, J.R. Troncoso-Pastoriza, K. Huguenin, J.-P. Hubaux, ORide: a privacy-preserving yet accountable ride-hailing service, in *Proceedings of the 26th USENIX Security Symposium* (2017), pp. 1235–1252
232. R. Lu, X. Lin, H.J. Zhu, P.H. Ho, X. Shen, ECPP: efficient conditional privacy preservation protocol for secure vehicular communications, in *Proceedings of the IEEE Conference on Computer Communications* (2008), pp. 1903–1911
233. J. Shao, X. Lin, R. Lu, C. Zuo, A threshold anonymous authentication protocol for VANETs. IEEE Trans. Veh. Technol. **65**(3), 1711–1720 (2015)

234. F. Bonomi, M. Mitzenmacher, R. Panigrahy, S. Singh, G. Varghese, Beyond Bloom filters: from approximate membership checks to approximate state machines, in *Proceedings of the ACM SIGCOMM Conference on Applications, Technologies, Architectures, and Protocols for Computer Communications*, vol. 36(4) (2006), pp. 315–326

235. M. Chiang, S. Ha, C.-L. I, F. Risso, T. Zhang, Clarifying fog computing and networking: 10 questions and answers. IEEE Commun. Mag. **55**(4), 18–20 (2017)

236. Y. Dodis, R. Ostrovsky, L. Reyzin, A. Smith, Fuzzy extractors: how to generate strong keys from biometrics and other noisy data, in *Proceedings of the International Conference on the Theory and Applications of Cryptographic Techniques, Advances in Cryptology* (2004), pp. 523–540

237. S. Nakamoto, Bitcoin: a peer-to-peer electronic cash system (2009). Available: https://bitcoin.org/bitcoin.pdf

238. C. Huang, R. Lu, X. Lin, X. Shen, Secure automated valet parking: a privacy-preserving reservation scheme for autonomous vehicles. IEEE Trans. Veh. Technol. **67**(11), 11169–11180 (2018)

239. M. Scott, MIRACL: Multi-precision integer and rational arithmetic C/C++ Library. Available: http://www.certivox.com/miracl

240. J. Ni, K. Zhang, X. Lin, H. Yang, X. Shen, AMA: anonymous mutual authentication with traceability in carpooling systems, in *Proceedings of the IEEE International Conference on Communications* (2016), pp. 1–6

241. R.R. Clewlow, G.S. Mishra, Disruptive transportation: the adoption, utilization, and impacts of ride-hailing in the United States, Institute of Transportation Studies, University of California, Davis (Research Report UCD-ITS-RR-17-07), 2017

242. Uber vs Lyft: a comprehensive comparison (2018). Available: https://www.ridester.com/uber-vs-lyft

243. 2017: the year the rideshare industry crushed the taxi (2017). Available: https://rideshareapps.com/2015-rideshare-infographic

244. J. Ni, K. Zhang, Y. Yu, X. Lin, X. Shen, Privacy-preserving smart parking navigation supporting efficient driving guidance retrieval. IEEE Trans. Veh. Technol. **67**(7), 6504–6517 (2018)

245. S. Basudan, X. Lin, K. Sankaranarayanan, A privacy-preserving vehicular crowdsensing based road surface condition monitoring system using fog computing. IEEE Internet Things J. **4**(3), 772–782 (2017)

246. https://www.uber.com (2018)

247. http://www.didichuxing.com (2018)

248. H. Yin, K. Gai, Z. Wang, A classification algorithm based on ensemble feature selections for imbalanced-class dataset, in *IEEE 2nd International Conference on Big Data Security on Cloud (BigDataSecurity)* (IEEE, Piscataway, 2016), pp. 245–249

249. Private mobility services need to share their data. Here's how (2017). Available: https://www.citylab.com/transportation/2017/07/private-mobility-services-need-to-share-their-data-heres-how/532482

250. M.H. Au, W. Susilo, Y. Mu, Constant-size dynamic k-TAA, in *Proceedings of the 5th International Conference on Security and Cryptography for Networks (SCN)* (2006), pp. 111–125

251. E. Ben-Sasson, A. Chiesa, C. Garman, M. Green, I. Miers, E. Tromer, M. Virza, Zerocash: decentralized anonymous payments from bitcoin, in *Proceedings of the IEEE 35th Symposium on Security and Privacy (S&P)* (2014), pp. 459–474

252. C. Garman, M. Green, I. Miers, Accountable privacy for decentralized anonymous payments, in *Proceedings of the 20th International Conference on Financial Cryptography and Data Security (FC)* (2016), pp. 81–98

253. S. Nakamoto, Bitcoin: a peer-to-peer electronic cash system (2008). Available: https://bitcoin.org/bitcoin.pdf

254. E. Androulaki, A. Barger, V. Bortnikov, C. Cachin, K. Christidis, A.D. Caro, D. Enyeart, C. Ferris, G. Laventman, Y. Manevich, S. Muralidharan, C. Murthy, B. Nguyen, M. Sethi, G. Singh, K. Smith, A. Sorniotti, C. Stathakopoulou, M. Vukolic, S.W. Cocco, J. Yellick, Hyperledger fabric: a distributed operating system for permissioned blockchains, in *Proceedings of the 13th European Conference on Computer Systems (EuroSys)* (2018), pp. 1–15

255. R. Li, A.X. Liu, Adaptively secure conjunctive query processing over encrypted data for cloud computing, in *Proceedings of the IEEE 33rd International Conference on Data Engineering* (2017), pp. 697–708

256. T.P. Pedersen, Non-interactive and information-theoretic secure verifiable secret sharing, in *Proceedings of the 11st Annual International Cryptology Conference (CRYPTO)* (1991), pp. 129–140

257. https://aws.amazon.com (2018)

258. H.T. Vo, A. Kundu, M. Mohania, Research directions in blockchain data management and analytics (2018). Available: https://openproceedings.org/2018/conf/edbt/paper-227.pdf

259. R. Bost, R.A. Popa, S. Tu, S. Goldwasserm, Machine learning classification over eEncrypted data, in *Proceedings of the Network and Distributed System Security (NDSS) Symposium* (2015), pp. 331–344

260. K. Gai, M. Qiu, S.A. Elnagdy, Security-aware information classifications using supervised learning for cloud-based cyber risk management in financial big data, in *IEEE 2nd International Conference on Big Data Security on Cloud (BigDataSecurity)* (IEEE, Piscataway, 2016), pp. 197–202

261. K. Gai, M. Qiu, Reinforcement learning-based content-centric services in mobile sensing. IEEE Netw. **32**(4), 34–39 (2018)

262. Y. Li, K. Liang, X. Tang, K. Gai, Waveband selection based feature extraction using genetic algorithm, in *2017 IEEE 4th International Conference on Cyber Security and Cloud Computing (CSCloud)* (IEEE, Piscataway, 2017), pp. 223–227

263. K. Gai, M. Qiu, Optimal resource allocation using reinforcement learning for IoT content-centric services. Appl. Soft Comput. **70**, 12–21 (2018)

264. S. Li, A. Leider, M. Qiu, K. Gai, M. Liu, Brain-based computer interfaces in virtual reality, in *2017 IEEE 4th International Conference on Cyber Security and Cloud Computing (CSCloud)* (IEEE, Piscataway, 2017), pp. 300–305

265. Z. Zhang, M. Li, L. Zhu, X. Li, SmartDetect: a smart detection scheme for malicious web shell codes via ensemble learning, in *Proceedings of the 3rd International Conference on Smart Computing and Communication*, Dec 2018, Best Paper Award

266. W. Liu, J. He, M. Li, R. Jin, Z. Zhang, An efficient supervised energy disaggregation scheme for power service in smart grid. Intell. Autom. Soft Comput. **PP**(99), 1–10 (2018)

267. L. Zhu, M. Li, Z. Zhang, X. Du, M. Guizani, Big data mining of users' energy consumption pattern in wireless smart grid. IEEE Wirel. Commun. **25**(1), 84–89 (2018)

268. S. Bi, R. Zhang, Z. Ding, S. Cui, Wireless communications in the era of big data. IEEE Commun. Mag. **53**(10), 190–199 (2015)

269. Y. Li, K. Gai, Z. Ming, H. Zhao, M. Qiu, Intercrossed access controls for secure financial services on multimedia big data in cloud systems. ACM Trans. Multimedia Comput. Commun. Appl. **12**(4s), 67 (2016)

270. M. Qiu, K. Gai, Z. Xiong, Privacy-preserving wireless communications using bipartite matching in social big data. Futur. Gener. Comput. Syst. **87**, 772–781 (2018)

271. H. Zhao, K. Gai, J. Li, X. He, Novel differential schema for high performance big data telehealth systems using pre-cache, in *IEEE 17th International Conference on High Performance Computing and Communications (HPCC)* (IEEE, Piscataway, 2015), pp. 1412–1417

272. H. Zhao, M. Qiu, M. Chen, K. Gai, Cost-aware optimal data allocations for multiple dimensional heterogeneous memories using dynamic programming in big data. J. Comput. Sci. **26**, 402–408 (2018)

273. M. Qiu, W. Dai, K. Gai, *Mobile Applications Development with Android: Technologies and Algorithms* (Chapman and Hall/CRC, Boca Raton, 2016)
274. Y. Li, K. Liang, X. Tang, K. Gai, Cloud-based adaptive particle swarm optimization for waveband selection in big data. J. Signal Process. Syst. **90**(8–9), 1105–1113 (2018)
275. A.L. Steenkamp, A. Alawdah, O. Almasri, K. Gai, N. Khattab, C. Swaby, R. Abaas, Teaching case enterprise architecture specification case study. J. Inf. Syst. Educ. **24**(2), 105 (2013)
276. R. DeStefano, L. Tao, K. Gai, Improving data governance in large organizations through ontology and linked data, in *2016 IEEE 3rd International Conference on Cyber Security and Cloud Computing (CSCloud)* (IEEE, Piscataway, 2016), pp. 279–284
277. S. Khan, Z. Zhang, L. Zhu, M. Li, Q.G.K. Safi, X. Chen, Accountable and transparent TLS certificate management: an alternate public-key infrastructure (PKI) with verifiable trusted parties. Secur. Commun. Netw. **2018**(8527010), 1–16 (2018)
278. A. Miller, A. Juels, E. Shi, B. Parno, J. Katz, Permacoin repurposing bitcoin work for data preservation, in *Proceedings of the IEEE Symposium on Security and Privacy (S&P)* (2014), pp. 475–490
279. A massive amount of storage sits unused in data centers and hard drives around the world (2019). Available: https://filecoin.io
280. R. Lu, X. Lin, Z. Shi, X. Shen, A lightweight conditional privacy-preservation protocol for vehicular traffic-monitoring systems. IEEE Intell. Syst. **28**(3), 62–65 (2013)
281. M. Sharples, J. Domingue, The blockchain and kudos: a distributed system for educational record, reputation and reward, in *Proceedings of the European Conference on Technology Enhanced Learning* (2016), pp. 490–496
282. X. Tang, C. Wang, X. Yuan, Q. Wang, Non-interactive privacy-preserving truth discovery in crowd sensing applications, in *IEEE Conference on Computer Communications (INFOCOM)* (2018), pp. 1–9
283. K. Gai, Z. Du, M. Qiu, H. Zhao, Efficiency-aware workload optimizations of heterogeneous cloud computing for capacity planning in financial industry, in *2015 IEEE 2nd International Conference on Cyber Security and Cloud Computing (CSCloud)* (IEEE, Piscataway, 2015), pp. 1–6
284. K. Gai, M. Qiu, H. Hassan, Secure cyber incident analytics framework using Monte Carlo simulations for financial cybersecurity insurance in cloud computing. Concurrency Comput. Pract. Experience **29**(7), e3856 (2017)
285. K. Gai, M. Qiu, M. Liu, Z. Xiong, In-memory big data analytics under space constraints using dynamic programming. Futur. Gener. Comput. Syst. **83**, 219–227 (2018)
286. H. Jean-Baptiste, M. Qiu, K. Gai, L. Tao, Meta meta-analytics for risk forecast using big data meta-regression in financial industry, in *2015 IEEE 2nd International Conference on Cyber Security and Cloud Computing (CSCloud)* (IEEE, Piscataway, 2015), pp. 272–277
287. S. Jayaraman, L. Tao, K. Gai, N. Jiang, Drug side effects data representation and full spectrum inferencing using knowledge graphs in intelligent telehealth, in *2016 IEEE 3rd International Conference on Cyber Security and Cloud Computing (CSCloud)* (IEEE, Piscataway, 2016), pp. 289–294
288. H. Jean-Baptiste, M. Qiu, K. Gai, L. Tao, Model risk management systems-back-end, middleware, front-end and analytics, in *2015 IEEE 2nd International Conference on Cyber Security and Cloud Computing (CSCloud)* (IEEE, Piscataway, 2015), pp. 312–316
289. X. Yu, T. Pei, K. Gai, L. Guo, Analysis on urban collective call behavior to earthquake, in *IEEE 17th International Conference on High Performance Computing and Communications (HPCC)* (IEEE, Piscataway, 2015), pp. 1302–1307
290. K. Gai, M. Qiu, M. Liu, H. Zhao, Smart resource allocation using reinforcement learning in content-centric cyber-physical systems, in *International Conference on Smart Computing and Communication* (Springer, Berlin, 2017), pp. 39–52
291. J. Dean, S. Ghemawat, MapReduce: simplified data processing on large clusters, in *Proceedings of the 11th USENIX Symposium on Operating System Design and Implementation (OSDI)* (2004), pp. 1–13

292. M. Qiu, K. Gai, B. Thuraisingham, L. Tao, H. Zhao, Proactive user-centric secure data scheme using attribute-based semantic access controls for mobile clouds in financial industry. Futur. Gener. Comput. Syst. **80**, 421–429, (2018)

293. A. Ouaddah, A.A. Elkalam, A.A. Ouahman, Towards a novel privacy-preserving access control model based on blockchain technology in IoT, in *Proceedings of the Europe and MENA Cooperation Advances in Information and Communication Technologies* (2017), pp. 523–533

294. Ethereum Smart Contract Best Practices - Known Attacks (2016). Available: https://consensys.github.io/smart-contract-best-practices/known_attacks

295. G. Ateniese, B. Magri, D. Venturi, E. Andrade, Redactable blockchain - or -rewriting history in bitcoin and friends, in *Proceedings of the 2nd IEEE European Symposium on Security and Privacy (EuroS&P)* (2017), pp. 1–9

296. ETHEREUM: a secure decentralised generalised transaction ledger EIP-150 revision. Available: https://gavwood.com/paper.pdf

297. Download Geth No Nick (v 1.8.19). Available: https://ethereum.github.io/go-ethereum/downloads

298. Ethereum Wallet and Mist Beta 0.11.1 - windows hotfix. Available: https://github.com/ethereum/mist/releases

299. K. Gai, M. Qiu, S.A. Elnagdy, A novel secure big data cyber incident analytics framework for cloud-based cybersecurity insurance, in *IEEE 2nd International Conference on Big Data Security on Cloud (BigDataSecurity)* (IEEE, Piscataway, 2016), pp. 171–176

300. G. Alipui, L. Tao, K. Gai, N. Jiang, Reducing complexity of diagnostic message pattern specification and recognition on in-bound data using semantic techniques, in *2016 IEEE 3rd International Conference on Cyber Security and Cloud Computing (CSCloud)* (IEEE, Piscataway, 2016), pp. 267–272

301. H. Jean-Baptiste, L. Tao, M. Qiu, K. Gai, Understanding model risk management-model rationalization in financial industry, in *2015 IEEE 2nd International Conference on Cyber Security and Cloud Computing (CSCloud)* (IEEE, Piscataway, 2015), pp. 301–306

302. C.A. Ardagna, M. Cremonini, E. Damiani, S.D.C. Di Vimercati, P. Samarati, Location privacy protection through obfuscation-based techniques, in *Proceedings of the 21st Annual IFIP WG 113 Working Conference on Data & Applications Security* (2007), pp. 47–60

303. M. Duckham, L. Kulik, A formal model of obfuscation and negotiation for location privacy, in *Proceedings of the 3rd International Conference on Pervasive Computing* (2005), pp. 153–170

304. T. Zhao et al., An anonymous payment system to protect the privacy of electric vehicles, in *Proceedings of the Wireless Communications and Signal Processing (WCSP)*, Dec 2014, pp. 1–6

305. M. Shen et al., Cloud-based approximate constrained shortest distance queries over encrypted graphs with privacy protection. IEEE Trans. Inf. Forensics Secur. **13**(4), 940–953 (2018)

Printed in the United States
By Bookmasters